THE EXPANSE OF HEAVEN

THE EXPANSE OF HEAVEN:

WHERE CREATION AND ASTRONOMY INTERSECT

Danny R. Faulkner

Master Books®

First printing: October 2017

Master Books®, P.O. Box 726, Green Forest, AR 72638
Master Books® is a division of the New Leaf Publishing Group, Inc.

ISBN: 978-1-68344-098-7
ISBN: 978-1-61458-633-3 (digital)
Library of Congress Number: 2017953955

Front and back cover photograph by Glen Fountain, Ex Nihilo Observatory. Used by permission.

Scripture quotations, unless otherwise noted, are taken from The Holy Bible, English Standard Version, copyright © 2001 by Crossway Bibles, a division of Good News Publishers. Used by permission. All rights reserved.

Scripture quotations so noted are taken from the New King James Version®. Copyright © 1982 by Thomas Nelson. Used by permission. All rights reserved.

Please consider requesting that a copy of this volume
be purchased by your local library system.

Printed in the United States of America

Please visit our website for other great titles:
www.masterbooks.com

For information regarding author interviews,
please contact the publicity department at (870) 438-5288.

Master
Books®
A Division of New Leaf Publishing Group
www.masterbooks.com

To my good friend and research partner for two decades,
Ron Samec.
Ron, your iron has sharpened my iron.

CONTENTS

FOREWORD

Astronomy questions continue to generate popular interest in the media: Is there alien life on the thousands of planets circling distant stars? Are we in imminent danger of an earth collision with a wandering asteroid or comet? Could the sun suddenly misbehave and thrust the earth into either a frozen or scorched future?

In contrast to these doubtful questions, positive refreshment is found in the biblical creation worldview of history and the future. Author Danny Faulkner provides such details in this astronomy summary with an expertise ranging from ancient cosmologies to gravity waves. The Creator's fingerprint is clearly evident throughout astronomy including the laws of nature. These laws describe gravity, conservation of energy, celestial mechanics and much more. The laws are unchanging and established for our survival and well-being. Physics and astronomy texts describe these laws but cannot explain their ultimate origin which lies in the realm of supernatural creation.

The field of astronomy is overwhelming to many of us. Our minds may "glaze over" when we consider light-year distances, vast spiral galaxies, and stars as numerous as the sand grains on all the seashores of the world. Author Danny Faulkner has a special ability for giving clear explanations of such details in his teaching and writing.

I am honored to serve with author Danny Faulkner on the board of directors of the Creation Research Society. This International Society has promoted creation studies including astronomy for over five decades. This new book adds to our mission of promoting quality biblical science.

DON DEYOUNG, PRESIDENT
CREATION RESEARCH SOCIETY
MAY 1, 2017

ACKNOWLEDGMENTS

As I finished the final edits of *The Created Cosmos: What the Bible Reveals About Astronomy* in the spring of 2016, I realized the need for this companion book. Once again, I rounded up the usual gang of suspects. Since Lee Anderson was so valuable in completing the earlier book, I knew that he would be most helpful in writing this one. Lee once again helped organize the material and acted as my editor. This new book required many more figures and illustrations than the previous book did; and Lee suggested many of the ones that we needed, secured the illustrations, and kept track of them during the editing and preparation stages. As before, I called upon my friend and astronomy colleague, Robert Hill, to review the manuscript's content. Bob, too, recommended many figures and illustrations included in this book. Once again, Steve Golden proofread the text. I'm amazed that the many little things that we missed—but these didn't get past Steve. This book is far better than it would have been without the tremendous help of these three gentlemen. I also want to thank my department director at Answers in Genesis, Andrew Snelling, for his support for this project. Being a geologist, Andrew also read the chapter on the planet earth, and he made many good recommendations for changes. Also, as before, I wish to thank Laurel Hemmings for her masterful work on the layout of this book.

I especially thank Don DeYoung, President of the Creation Research Society, for kindly agreeing to write the foreword. Don has written multiple works on creation and astronomy, such as *Astronomy and the Bible: Questions and Answers* and, along with John Whitcomb, *Our Created Moon: Earth's Fascinating Neighbor*. I've known Don for 25 years, and it has been my pleasure to serve with him on the board of directors of the Creation Research Society for more than a decade.

Finally, I thank my loving wife, Lynette, for her encouragement in completing this book.

DANNY R FAULKNER
APRIL 11, 2017

INTRODUCTION

I intend this book as a companion to *The Created Cosmos: What the Bible Reveals about Astronomy*. In the introduction to that book, I noted how it was different from other books about astronomy written by biblical creationists in recent years by comparing the difference between biblical theology and systematic theology. I likened the approach taken in *The Created Cosmos* to the approach of biblical theology, because in it I explored what the Bible had to say about astronomy, using the categories that the text itself presents, and discussing topics in terms of how the original readers of Scripture would have understood them. (I also included a discussion of a few particularly relevant things that the Bible *doesn't* say about astronomy.) A few people may have been frustrated with that book. There are two possible reasons for this frustration. One reason is that *The Created Cosmos* was not explicitly oriented toward recent, six-day creation. Recent creation was discussed, or otherwise there would have been no point in including the chapter on the light travel time problem. Other than that, though, many of the chapters did not address the age of creation at all. However, keep in mind that the direct context of most of the biblical passages discussed in that book did not specifically address the age of creation, so it would have been out of place to introduce age.

The present book is more of a systematic treatment. Its purpose is to discuss astronomy, using scientific categories, in the light of Scripture. Since I clearly see the Bible as teaching that God created the world in only six days, and that the creation was only thousands of years ago, this will be an underlying assumption in this book. I will describe various aspects of astronomy as it relates to that understanding. This is sure to satisfy anyone who may have been disappointed by what they might have perceived as a lack of sufficient attention to recent, six-day creation in *The Created Cosmos*.

The second reason some people may have been frustrated with *The Created Cosmos* was the many astronomical topics that it omitted. This present book contains far more astronomy than *The Created Cosmos* did. However, it is not meant to be exhaustive. Inevitably, some interesting astronomical topics will not be discussed. I endeavored to include material relevant to the biblical creation model of astronomy. Many astronomical topics may be of interest, but they may not have any relevance to biblical creation. As such, this book is not intended as a textbook on astronomy, though it could be so used, and it certainly could be used as a supplement for a course on astronomy.

For a long time, I have considered Paul Steidl's excellent 1979 book *The Earth, the Stars, and the Bible* to be the best book written thus far on creation astronomy.[1] It is well-written, and it covers many relevant topics. However, being nearly 40 years old, the book is a bit out of date. In many respects, I hope the book that you are now reading will replace and perhaps exceed that book.

Why Study Astronomy?

I may be biased, but I have always thought that astronomy is the most fascinating science. Many people simply are not that interested in science in general. Perhaps it came across as boring in school. Or maybe science is perceived as geeky, or a subject that only very bright people can understand. It is a pity that our educational system can turn off countless people to a variety of fascinating subjects simply because the way that we educate succeeds in sucking the life out of so many things. However, most people seem to be fascinated with astronomy. There is wonder to the natural world, wonder that I think at least hints at the wonder of the Creator. Most people would agree that astronomy is packed full of wonder. Part of the wonder of astronomy is the vastness of the universe. Douglas Adams, in his best-selling book *The Hitchhiker's Guide to the Galaxy* said it very well:

> Space is big. Really big. You just won't believe how vastly hugely mindbogglingly big it is. I mean you may think it's a long way down the road to the chemist's,[2] but that's just peanuts to space.[3]

---

[1] Paul Steidl, *The Earth, the Stars, and the Bible* (Phillipsburg, New Jersey: Presbyterian and Reformed Publishing, 1979).
[2] Adams was English. In England, a chemist is what we in the United States would call a pharmacist.
[3] *The More than Complete Hitchhiker's Guide* (New York: Bonanza Books, 1989), 53.

To get across some of the distances involved, when teaching astronomy at the university, I had my students work out a scale model of the solar system. In this model, the sun was about the size of a basketball, and the earth was the size of a BB nearly 100 ft (30 m) away from the basketball. Jupiter was approximately an inch across and 500 ft (150 m) from the basketball-sized sun. Neptune, the most distant planet from the sun, was a little more than a third of an inch (25 mm) across and 3000 ft (900 m) from the basketball-sized sun. If we were to place this model of the solar system in New York City, Alpha Centauri, the nearest star to the sun, would be a beach ball in Hawaii. In this scale model, most of the stars that you see at night would be far larger than a beach ball, but they would not fit on the earth because their distances would exceed the earth's circumference. But these are some of our nearest astronomical neighbors. The truly distant objects are so far away that the distance is difficult to comprehend.

Another part of the wonder of astronomy is the beauty that we see. How can anyone not be awed when looking at photographs of astronomical bodies, or better yet, seeing them directly? Both the vastness and the beauty ought to convince all that there is a God. The Bible relates the wonder of astronomy to our Creator. Psalm 19:1 tells us that the heavens declare God's glory, and that God has made everything. In Psalm 8:3–8, gazing at the night sky inspired David to reflect upon man's place in the world. In Job 9:8–9, Job referenced astronomical bodies in contemplating the power of God, while God in turn challenged Job's thinking (Job 38:31–32) by making reference to various astronomical bodies. Given the God-honoring nature of astronomy, the study of astronomy ought to be the burning desire of every Christian. But then, perhaps I am biased.

Astronomy is different from other sciences. One large difference is that astronomy is almost entirely observational rather than experimental science. With other sciences, we use our hypotheses and theories to make predictions concerning the outcomes of experiments and then test our predictions by conducting those experiments. However, in astronomy we are so far removed physically from the subjects of our study that we generally cannot conduct experiments.[4] Instead, we make predictions of what certain observations might reveal and then carry out those observations to test our predictions. In this manner, astronomy is more passive than other sciences.

[4] Exceptions would include studies of samples from astronomical bodies. At this time, the only samples that we have are meteorites, lunar rocks returned by the Apollo astronauts, and microscopic pieces of a comet returned to earth.

The Assumption of Naturalism

Science often is defined as the study of the natural world using the five senses. Notice that since the subject of science is the natural world, there is much emphasis on nature in science. Indeed, what we now call *science* originally was called *natural philosophy*. It was William Whewell in the 1830s who suggested the name change from *natural philosophy* to *science*, and that convention rapidly took hold. Previously, *science* referred to systematized study of any subject that was not considered an art.

Science largely consists of describing natural phenomenon in terms of consistent patterns that we observe. Consider physics, the most quantitative of the sciences. We see regularity in the way the physical world works. Physicists describe this behavior in terms of mathematical equations that can be relatively simple but always elegant. Scientists often revel in the patterns that we see in nature, but far too few of them dwell on the question of *why* those patterns are there. It generally is assumed that the patterns simply exist with no more thought given to the matter. This seeming lack of interest in the source or origin of these patterns in nature easily and subtly can shift into the assumption of *naturalism*, the belief that the natural world is all that exists. A scientist with this persuasion understandably may object that the question of the source of the patterns in nature is beyond the scope of science, as far as the practice of science generally is concerned today. The origin of the patterns in nature is a philosophical question rather than a scientific one. Perhaps when science was known as natural philosophy, this question would have been more welcome than it is today.

All people, including scientists, begin with certain assumptions, and the answer to the philosophical question of the source of patterns in nature falls under the category of assumptions. Many scientists of the past and present see the handiwork of God in the patterns and regularity of nature. This is consistent with Colossians 1:17, which states that in Christ all things hold together, and Hebrews 1:3, which teaches that Christ upholds all things by the power of His word. If the Creator sustains the world moment by moment, one might expect this to be done in a consistent manner which we then perceive as pattern. Again, a scientist committed to naturalism may object that this is a metaphysical assertion because belief in a Creator is an assumption of a spiritual reality beyond the physical. Indeed, this is a metaphysical assertion, but is this a fair criticism? The assertion that there is no Creator equally is a metaphysical assertion. While a scientist committed to naturalism may sincerely believe that only his assumption is logical and proper, it is an assumption and hence cannot be demonstrated. Therefore, neither that assumption of naturalism nor its denial is superior or inferior, but rather they are opposites that are equivalent in terms of proper argumentation.

Does it matter whether one assumes naturalism or not when generally practicing science? When practicing what some people call operational science or experimental/observational science, no, because we can agree that pattern and regularity exist apart from addressing any reason as to why they exist. However, some modern scientists have altered the definition of science from being *the study of the natural world using the five senses* to being *the search for natural explanations*. This new definition is a very subtle yet profound shift that tacitly embraces naturalism. It specifically excludes the possibility that a Creator occasionally has or might again interact with His creation. Those who believe in biblical creation recognize that God normally sustains the creation and any deviation from that would be a rare event, an event that few scientists would ever encounter. Therefore, in the matter of operational science, there does not appear to be much harm in accepting this new definition of science. However, this new definition precludes God from *ever* involving Himself in the operation of the world. More specifically, this new definition denies that God created the world. Note that this denial of creation cannot be demonstrated, but simply is asserted.

This new philosophy of science has become so deeply entrenched in the minds of many people to the extent that it appears intuitively obvious to them that the world must have a natural explanation. Thus, atheism has come to dominate the direction of what some call historical, or origins, science. Historical science is the attempt to elucidate processes that occurred in the past. Since these are past processes, they cannot be tested today in the traditional way that science works. Hence, the rules of historical science are a bit different from those of operational science. One may attempt to reproduce the conditions of the past, but how do we know what those conditions were? We must assume those conditions; but if the conditions were different from what we assume, then the conjectured processes likely are in error. For instance, there is no single, widely accepted theory of how Grand Canyon formed, even among those who are committed to naturalism. Rather, there are many different theories. Each theory can explain certain aspects of Grand Canyon reasonably well, but cannot explain other aspects. If historical science were such a straightforward enterprise as many proponents would have you believe, why are there so many different theories? Unlike operational science, we cannot perform a laboratory experiment or an observation that would settle the issue.

As you might expect, there are many theories dealing with origins in astronomy. Probably the ultimate evolutionary theory is the origin of the universe. However, what if we apply the first and second laws of thermodynamics to the history of the universe? The first law of thermodynamics is the conservation of energy—energy can neither be created nor destroyed.[5] If the universe had a beginning, then that would amount to the sudden appearance of energy, and thus it would violate this law of conservation. Therefore, the first law of thermodynamics would seem to preclude energy and the universe having a beginning. If there was no beginning, then energy must be eternal. However, the second law of thermodynamics dictates that while the total energy of the universe remains constant, the energy available to do work decreases with time. Physicists quantify this with entropy, defined so that it always increases as the available energy decreases. Therefore, entropy *always* increases. If the universe is eternal, then more than enough time has elapsed for the universe to have reached maximum entropy, with no useful energy remaining. This clearly has not happened, so the universe cannot be eternal.

The first and second law of thermodynamics are two of the most tested theories in science. They appear to apply universally, and they clearly do so today. Applying the assumption of naturalism, the two laws must have applied indefinitely into the past. But the first law requires that the universe be eternal, while the second law requires that the universe have a finite age. This is a contradiction. How can we resolve this dilemma? A possibility is that one or both of the first and second laws of thermodynamics must not have applied universally in the past, but how do you know which? Furthermore, altering either one in the past would violate the assumption of universality of physical law throughout space and time, an assumption upon which science is based and thus

---

[5] The first law of thermodynamics was formulated before the modern understanding of the equivalence of matter and energy via Albert Einstein's famous equation, $E = mc^2$. We now recognize that it is the sum of matter and energy that is conserved, though in most instances we merely speak of energy.

every scientist must make. This would be a huge price to pay. Once one abandons the universality of physical law, then one cannot have any real confidence that universality will hold in today's world, rendering science impossible.

But let us suppose for the sake of argument that sometime in the past one of these two laws was violated. For instance, if the universe suddenly appeared, that would constitute a huge violation of the first law of thermodynamics. How could we apply a scientific test to determine if this indeed happened? We cannot, which illustrates my point that we do not test theories of historical science the same way that we test theories of operational science. More importantly, since this would have been a radical departure from the manner in which the world normally works, would this not constitute a miracle? A person committed to naturalism may assert that this violation is no violation at all but merely some not yet understood natural mechanism.[6] However, this is grasping at straws. When confronted with what appears to be a genuine miracle, a person committed to naturalism retreats into special pleading to preserve his worldview. This underscores the perverse and pervasive nature of the assumption of naturalism. Naturalism cannot allow any departure from naturalistic explanations, to the extent that any departures are excluded more or less by definition. The takeaway from this discussion is that the question of ultimate origins is a metaphysical one.

Once one recognizes the metaphysical nature of the question of origins, it does not necessarily follow that the God of the Bible is the Creator. However, it does follow that the God of the Bible is a viable answer to the question of ultimate origins. It is the assumption of this book that He is, and I hope that it also is the assumption of you, the reader. But even that is not the most important point. You may agree with everything that I write about astronomy, but it is all for naught if you do not have a personal relationship with Jesus Christ, our Creator and Redeemer (John 1:1–18). I hope that this book will encourage the Christian in his faith and that it will help the lost to find salvation through the finished work of Jesus Christ.

---

[6] A good example of this is Lawrence M. Krauss' book *A Universe from Nothing: Why There Is Something Rather Than Nothing* (New York, New York: Free Press, 2012). Krauss used a questionable interpretation of quantum mechanics, along with many speculative ideas about physics, to argue that we have good physical reasons to believe that the universe came into existence by itself. While exuding confidence throughout, Krauss' argument amounts to a faith-statement about his own reasoning, undergirded by the metaphysical assumption of naturalism. I will discuss this further in Chapter 9.

CHAPTER 1

A Historical Perspective:
Ancient Cosmologies and the Bible

The Ancient Near East and the Bible's Cosmology

The Bible largely is a product of the ancient Near East (ANE). As such, it has become fashionable today to study and interpret the Bible in terms of its ANE background. This goal is laudable, even necessary, for some things in Scripture involving cultural context are difficult to understand fully in our modern world. There are certain customs observed in the Bible which are foreign to our modern contexts. For instance, marriage customs in ancient Israel were different from ours today, so the account of Joseph's intended response to his betrothed Mary's pregnancy in Matthew 1:18–19 is difficult for us to comprehend without more information. To twenty-first century people reading the Bible, subtlety and fine detail often is lost, and it is easy to misinterpret the author's meaning. However, it is not proper to use the ANE background to reinterpret passages to mean something other than what the author clearly intended. A prime example of how authorial intent is violated is to claim that the ANE background of the Bible's creation narrative provides justification for reinterpreting the plain sense of what that narrative expresses, especially as it concerns the temporal aspects of God's creative work.

One approach in reinterpreting the creation account this way is to note that the creation account has certain aspects of a polemic against the pagan gods of the cultures surrounding ancient Israel. For instance, the sun and moon were deities worshiped by nearby cultures, but the Genesis 1 creation account does not even mention the sun and moon by name. Rather it refers to them as the greater and lesser lights. The purpose for this probably was twofold.

First, the creation account does not even deign to mention the names of those pagan gods. Second, the true and living God made the sun, moon, and stars. Because God made the sun, moon, and stars as non-sentient and non-living beings, they are not gods. While Genesis 1:1–2:3 functions as a polemic on a larger scale to communicate the fact that God is unopposed in His creation of the world, it does not follow that the text is merely a polemic. The major point of chapter 1, that the Lord God created the universe through sovereign, unopposed action, is part and parcel to the polemic. However, the main point of the chapter is not the polemic itself, but the theological truth conveyed in the polemic. Yet some have seriously suggested that the details of Genesis 1 do not matter, but instead it is some subtle message behind the creation account that matters. The truly important message supposedly is that God created the world, but that the exact details are inconsequential. Related to this approach is the claim that the creation account of Genesis 1:1–2:3 is a poetic description of the creation event rather than historical narrative. There indeed are some poetic elements present, but that does not make them ahistorical. Notably, there is an unhealthy trichotomy that has emerged even in conservative studies, which drives a wedge between the study of the text as literature, as history, and as theology. The text is *all three of these*, and these emphases must be studied in unison.

Playing off this false trichotomy, some claim that while we today are understandably confused into thinking that this is historical narrative, if we considered the ANE context, we would realize that the ancient Hebrews knew better. However, within the context of ANE culture, what would the ancient Hebrews have thought? They were surrounded by cultures that each had their own respective creation stories. Certainly, the surrounding nations thought that their stories reflected truth; otherwise, why did they persist in believing something that they knew to be false? If the ancient Hebrews thought about their creation story any differently, then they were not thinking in terms of the ANE background. What of the polemic elements of Genesis 1:1–2:3? A polemic is supposed to be a refutation of an idea. If the ancient Hebrews believed that the creation account in Genesis was not true, then it amounted to myth. One cannot legitimately refute one mythology with another mythology. Only truth can properly refute myths.

A different approach to reinterpreting the first few chapters of Genesis in accordance with other ANE literature is to claim that the first part of Genesis is an amalgamation of stories prevalent among societies surrounding the Hebrews in ancient times. Notice that this approach is completely opposite to the approach previously discussed, so both clearly cannot be true at the same time. In this view, the creation account contains aspects of cosmology that commonly were believed in the ANE, but now are known to be false. But the Bible allegedly is unscathed by this, because the details of the creation account do not matter. One must instead look at the big picture. According to this theory, the real "truth" behind the story somehow was lost, so the belief that the creation account was historical narrative eventually took hold and managed to become emplaced as truth, particularly in the West because of the influence of Christianity.

This view of history is illustrated by the engraving reproduced in Figure 1.1. This illustration, in turn, perpetuates this view of history. The engraving

Figure 1.1. Flammarion engraving. (Public Domain).

shows a flat earth with an edge to the left. On either side are pillars that support heaven, which is on the top and to the left. In between is the domed vault of the sky, to which the sun, moon, and stars are attached. On the left, one can see an intrepid individual finding the edge of the vaulted dome and daring to peek through to see what lies beyond. The general public is familiar with this engraving, because it has been reproduced countless times. It is commonly believed that the engraving is of medieval origin, but the general public is wrong. The engraving first appeared in Camille Flammarion's 1888 book, *L'atmosphère: Météorologie Populaire* (*The Atmosphere: Popular Meteorology*). Hence, rather than being a product of the Middle Ages, this engraving is the product of the late nineteenth century.

More than anything, the Flammarion engraving, as it is known, illustrates the hatchet job that the so-called Enlightenment did on the Middle Ages, Christianity, and the Bible. It was Enlightenment thinkers who coined the term "dark ages" to refer to the Middle Ages, though the Middle Ages were not nearly as dark as many of us were taught in school. The Enlightenment of the eighteenth century laid the foundation for higher criticism, uniformitarianism, and Darwinism of the nineteenth century. By the time of the publication of Flammarion's book in the late nineteenth century, the damage to the reputation of the Middle Ages had already been done. But the real objective was to discredit the Bible. The storyline that had developed was that the Bible's cosmology was fundamentally wrong, but that its cosmology had nevertheless become entrenched in the West due to the influence of Christianity and, accordingly, that it remained the status quo until modern times. As the argument goes, it was time for modern scientific thinking to unseat antiquated ideas. Let us now begin evaluating this supposed history.

What Shape Is the Earth?

It is common knowledge today that until about five centuries ago, nearly everyone thought that the earth was flat. However, common knowledge often is wrong, as it is in this case.[1] The earliest record that we have of belief in a spherical earth is that of Pythagoras, in the late sixth century BC. Most people today find this difficult to believe, because it so contradicts what they have heard

[1] An excellent discussion of how the late ancient and medieval world understood the earth is spherical is found in Russell, J. B. 1991. *Inventing the Flat Earth: Columbus and Modern Historians*. New York, New York: Praeger.

their entire lives. People recognize that with modern technology the shape of the earth must be easy to determine, but how did supposedly primitive people in the past discern this? It really is quite simple. A lunar eclipse occurs when the earth passes exactly between the sun and moon so that earth's shadow falls on the moon. As you might expect, a lunar eclipse is not a common event, on average happening for any given location on earth less often than once per year. After observing a few lunar eclipses, one comes to realize that the earth's shadow cast on the moon always is round. If the earth were flat and round, like a disk, it could cast a circular shadow, but only during an eclipse that happened near midnight (Figure 1.2). An eclipse that happened near sunrise or sunset would cast a very differently shaped shadow. However, the earth's shadow always is circular, regardless of the time, and hence, regardless of the orientation of the earth. The only shape that always casts a circular shadow is a sphere. Therefore, the earth must be a sphere.

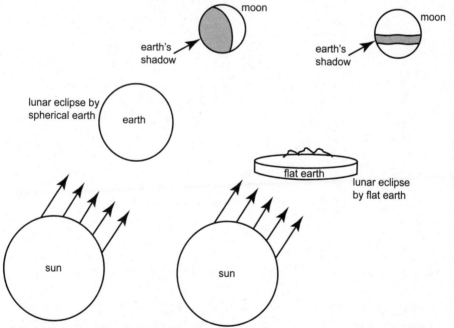

Figure 1.2. Hypothetical lunar eclipse by a flat earth.

There are other arguments for the earth's spherical shape of which the ancients were also aware. As one travels northward, the stars visible in the northern part of the sky rise higher, while stars visible in the south fall lower, even falling into invisibility below the horizon. Traveling southward, the effect is reversed—stars visible in the northern part of the sky descend and in some cases disappear, while stars in the south climb higher in the sky. For instance, from where I used to live in South Carolina, I could see Canopus, the second brightest-appearing star to us, low in the sky briefly on winter evenings. However, where I now live in northern Kentucky, at more than 4° farther north latitude, I cannot see Canopus at all. This can happen only if the earth is curved in the north-south direction (Figure 1.3).

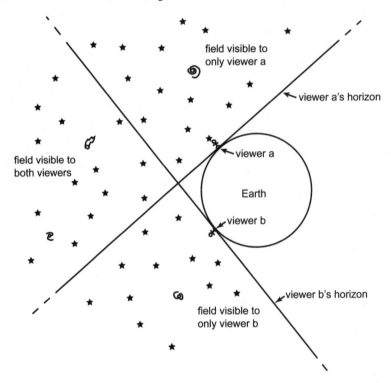

Figure 1.3. Different fields of view in the night sky (not to scale).

Related to this phenomenon was a legend that the Phoenicians circumnavigated Africa in early antiquity. The Phoenicians were excellent sailors, and in their time their culture dominated maritime trading in the Mediterranean world. The peak of their culture is conventionally dated between 1200 BC and 800 BC. Part of this legend states that on the other side of Africa, the sun was in the northern part of the sky. In the north temperate latitudes, the sun always is in the southern sky, but in southern temperate latitude, the sun always is in the northern sky. But this might not be expected, if the earth were flat. By the classical Greek period, the fifth and fourth centuries BC, the Greeks definitely knew that the earth was spherical, and they believed the Phoenician legend to be true, because of this one detail.

As for curvature in the east-west direction, a similar thing happens. You may be aware that the sun rises and sets about three hours earlier on the East Coast than on the West Coast of the United States because of the earth's spherical shape. With today's rapid transportation, instant communication, and accurate time pieces, this is easy to verify; but this was not possible in the ancient world. However, they had a different method. A lunar eclipse that began shortly after sunset at one location in the east would begin before sunset at a location that was many miles to the west. The eclipse began at the same instant, but at different local times due to the earth's curved surface. Of course, a single person in the ancient world could not observe this, but two people separated by a large distance could observe it and later compare their observations. This sort of experiment was noted in records from the ancient world. If the earth is curved both in the north-south and east-west directions, then its most likely shape is a sphere.

The ancients also noted that the hull of a departing ship would disappear before the mast would. This could happen only if the earth were spherical so that a ship sailing away passed below the visible horizon. Actually observing this is difficult without optical aid, and the telescope was not invented until four centuries ago. Another difficulty is atmospheric refraction that frequently can bend light around the curvature of the earth. More likely, the ancients observed this phenomenon in reverse. As a ship approaches land, the land first is visible to someone atop the mast. People on the deck cannot see the land yet, because the land lies below the horizon from their perspective. However, as the ship ventures closer to land, the land eventually becomes visible to those people on the deck. This can happen only if the earth is spherical.

These arguments are rigorous, and they established the earth's spherical shape among ancient people. Once established, people also developed other, less rigorous arguments for the earth's spherical shape. The sun and moon clearly are circular, because they appear round. However, are they round and flat in two dimensions, like a disk? Or are they round in three dimensions, like a sphere?

The sun constantly appears round, so if it is a disk rather than a sphere, then it must keep the plane of its disk perpendicular to our line of sight at all times. Likewise for the moon, except that the moon's phases offer a definitive clue to its three dimensional shape. Lunar phases are the expression of how much of the moon's surface visible to us is lit by the sun, and that in turn is the result of the relative geometry of the earth, moon, and sun.[2] The terminator is the division between light and dark on the moon. The terminator generally is curved. But that can be only if the moon is curved in a third dimension, along our line of sight. Therefore, the moon must be spherical. By analogy, if the earth is anything like the moon or the sun, it must be spherical too. This is proof by analogy, which is not as rigorous as the other methods.

Another less rigorous argument involved concepts of perfection. The ancient Greeks, and probably other ancient cultures, thought that the circle was the most perfect shape. A sphere merely is a circle spun in a third dimension. In their cosmology, it was fitting that the heavenly bodies were perfect, so it was not surprising that the sun and moon were spherical. One might expect that the earth too is spherical. However, this aspect was at odds with other characteristics of their cosmology. The earth, unlike the heavenly realm, was imperfect, so one could make the case that the earth did not follow this rule. Apparently, the ancient Greeks did not notice this possible inconsistency.

About 200 BC, Eratosthenes, a Greek living in Egypt, built on knowledge that the earth was spherical and measured accurately the size of the earth. He did this by measuring the difference in the altitude of the sun at noon at two locations widely separated by latitude on the summer solstice (Figure 1.4). Obviously, he

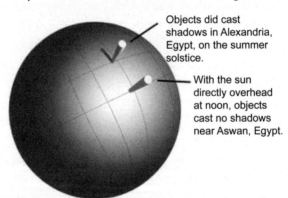

Objects did cast shadows in Alexandria, Egypt, on the summer solstice.

With the sun directly overhead at noon, objects cast no shadows near Aswan, Egypt.

Figure 1.4. Illustration of Eratosthenes' experiment (Danny R. Faulkner, *Universe by Design* [Green Forest, Arkansas: Master Books, 2004], 9).

[2] See Chapter 3 for further discussion of lunar phases.

could not do this the same year, but rather he made the measurements on the summer solstice two different years. The two locations were separated in the north-south direction by 500 mi (800 km). Eratosthenes found that the difference in the altitude of the sun was $1/50$ of a circle. Therefore, the circumference of the earth was 50×500 mi (800 km), or 25,000 mi (40,200 km). Of course, Eratosthenes used units other than miles, but the results that he got closely matched the modern value. Eratosthenes' measurement remained the standard size given for the earth for nearly two millennia.

By the time of Eratosthenes, belief in a spherical earth had been common since the time of Pythagoras, about three centuries earlier. Virtually all ancient Greek scientists and philosophers that we know of believed in a spherical earth. Plato and Aristotle clearly did, and their teachings played a crucial role in shaping the thinking of those in the Roman Empire and the West for two millennia after they lived. One is hard pressed to find any writings from the past 25,000 years that espoused a flat earth. So why do so many people today think that belief in a flat earth was prevalent until recently? The answer is the development of the *conflict thesis* in the late nineteenth century.[3]

The conflict thesis was a result of the growing hostility toward Christianity in the post-Enlightenment world. According to the conflict thesis, religion in general, and Christianity in particular, opposed thought and learning. Supposedly, it was not until man threw off the restraints of ignorance brought about by religion that intellectual pursuits could flourish. In this view, it was the shift from theism to humanism that enabled the beginning of science as we know it in the seventeenth century. In truth, it was the Protestant worldview of northern Europe that provided the fertile ground for science as we know it to develop. Not only did the humanists of the so-called Enlightenment hijack science, but they rewrote history so that they took credit for the development of science.

When Charles Darwin published his *Origin of Species* in 1859, much of the church opposed Darwin's ideas. The prime objective of the conflict thesis was to intimidate the church into giving up this fight.[4] The conflict thesis buried

---

[3] Key figures in the development of the conflict thesis were John William Draper (1811–1882) and Andrew Dickson White (1832–1918). Draper published *History of the Conflict between Religion and Science* in 1874, while White published *A History of the Warfare of Science with Theology in Christendom* in 1896.

[4] At the same time, a very different sort of battle was shaping up within academia concerning textual study. Higher criticism was attempting to argue for internal contradictions in the Bible (and rewriting the history of textual composition in the process). Hence, there was no one cause that led to the advancement of the secular worldview. It was the result of influences in multiple sectors. The conflict thesis was just one theater in this war of worldviews.

the church under a load of false guilt for its supposed transgression of resisting science and progress in the past. Exhibit A in this argument was the church's alleged opposition to the idea of a spherical earth. The church faced the charge that it had persecuted those who dared question whether the earth was flat. The story of Christopher Columbus was retold in a way to bolster this false history, and this false narrative unfortunately is deeply rooted in our belief system today. This story is so compatible with the false notion of the dark ages from which mankind slowly emerged during the Renaissance that it seems plausible to most people. However, even casual study of true history, or brief reflection on this myth, reveals how poorly founded this thesis is. For instance, Columbus sailed westward from Spain to the Caribbean and then sailed eastward back to Spain. Four times. How did that prove that the earth was spherical? That key part of the story doesn't even make sense, but it escapes the notice of most people today. While the conflict thesis was widely embraced by historians in the latter part of the nineteenth century and much of the twentieth century, it now is rejected by most historians. However, the conflict thesis remains deeply embedded in the belief system of most people today, and thus amounts to a "modern mythology."

To illustrate the lengths taken to rewrite history on this matter, consider medieval discussions of the *antipodes*. Antipodes are points on the earth that are diametrically opposite one another. The word *antipode* derives from Latin literally meaning "opposite feet," so the derivation of the word comes from the concept of a point on the opposite side of the earth, opposing where we stand (where our feet are planted). Knowing that the earth was spherical, Plato used an equivalent Greek term in his dialogue, *Timaeus*, as did Aristotle, Strabo, and Plutarch. There developed a belief that there was another large land mass on the other side of the earth opposite the Mediterranean world that possibly could be inhabited. However, the two were separated by a vast ocean, and transit between the two would require passage through the tropics. There was considerable discussion as to whether men could survive a voyage through the heat of the tropics.

By the Middle Ages, the antipodes came to refer less to this hypothetical land and more to possible inhabitants of that land. This use is traced to Isidore of Seville's early seventh century work *Etymologiae*. Many people in the Middle Ages came to doubt whether these people, or antipodes, existed. The difficulty or impossibility of transit to that far away land came to involve two related theological issues. First, if that land were inhabited, how did people, being descendants of Adam, arrive there? Augustine raised that objection. Second, the Great Commission commands that we preach the gospel to the entire world (cf. Matthew 28:18–20), but how could Christ's followers do this if transportation between the two inhabited parts of the earth was not possible? Further complicating this matter was an eighth century dispute between Boniface and Vergilius of Salzburg that eventually involved Pope Zachary. Some of the words of this dispute and later discussions expressing doubts about the antipodes have been misconstrued to refer to the opposite side of a spherical earth rather than to possible inhabitants on the other side of the earth. Hence, people who perpetuate the flat earth myth often use medieval quotes denying the existence of the antipodes as evidence of a denial of a spherical earth— when in fact the question addressed by those quotes was whether the opposite side of the world was inhabited.

Keep in mind that Pythagoras' understanding that the earth is spherical is only the first *recorded* mention; it is possible that some belief in a spherical earth predated Pythagoras. No written works of Pythagoras survive, so what we know of Pythagoras and his teachings relies upon much later sources. This is the case of the writings of people from this time period (late sixth century BC) and earlier. Early in the sixth century, Anaximander (who may have been Pythagoras' teacher) taught that the earth was a flat disk, with its diameter three times its thickness. Notice that in this model the earth is round, yet flat. This apparently was a common belief in ancient flat earth cosmologies. In Anaximander's cosmology, the Mediterranean world was the center of earth's land, hence the name for the Mediterranean Sea, which means "middle of the earth." That land was supposedly surrounded by a vast ocean that extended to the edge of the flat earth. This, too, was a common element of ancient flat earth cosmologies. However, while in most cosmologies the flat earth rested on other things, such as the backs of turtles or elephants, in Anaximander's view, the earth rested upon nothing. It is believed that most earlier Greek, Egyptian, and Mesopotamian cosmologies consisted of a flat disk, with central land surrounded by water. Ancient cosmologies of the Far East and of the Americas were similar, and they persisted into relatively modern times.

Contrast the common ancient belief that the earth rested upon something with Job 26:7, which reads,

He stretches out the north over the void and hangs the earth upon nothing.

The content of the book of Job is very old,[5] yet this statement does not conform to most of the cosmologies of the ANE, which typically assert that the earth rests upon something. If the Scriptures reflected the cosmologies of surrounding cultures, why is it different from them in this one important aspect? Some Christians have taken Isaiah 40:22, which speaks of the circle of the earth, as evidence that the Bible teaches a spherical earth. However, one ought to be circumspect here, because it is not clear that this verse actually means that the earth is spherical. In similar manner, skeptics have claimed that biblical passages that refer to the four corners of the earth (e.g., Isaiah 11:12) imply that the earth is flat. However, if these verses were meant to reflect common ANE cosmological beliefs, the earth could not have corners, because in most ANE cosmologies, the earth was flat, but round. The Hebrew word used here refers to extremity, and is translated thus elsewhere, such as in Numbers 15:38, where it is rendered "borders." From the Bible, the phrases "four corners of the earth" and "ends of the earth" (e.g., Job 37:3; 38:13) have entered the English language as idioms referring to the greatest extent of anything. We use these phrases today, though very few, if any, of the people using such phrases think that the earth is flat.

As of the time of the writing of this book, there has been a surge of interest in the idea that the earth is flat. Much of this has been promoted on social media, including many videos on various internet sites, such as YouTube. Unfortunately, a large number of Christians have been taken in by this idea. This topic is discussed in Appendix A.

Biblical Cosmology

Many cosmologies of the ANE were related intimately to cosmogonies, or, more properly, to *theogonies*. Theogony refers to the origin or genealogy of gods. Many of these cosmogonies amount to stories of how their gods came to be, with gods giving rise to other deities. Along the way, the earth and sky often were formed from parts of those gods' bodies. The sky and celestial objects frequently are manifestations of those gods. Many of those worlds arise out of primordial water, water that is sometimes eternal.

---

[5] The events of the book of Job predate the writing of the Pentateuch in the fifteenth century BC. It is not clear whether the book was written prior to Moses writing the Pentateuch or later.

Contrast this to the biblical creation account. There is only one self-existent eternal God, so there is no theogony. Matter is not eternal, but instead matter is a creation of God in the finite past. In the biblical creation account, the earth and sky are not body parts of a god or gods. The earth and sky are creations of God, but with no explicit statements concerning cosmology. Rather, to reach conclusions about biblical cosmology, one must draw inferences from certain biblical passages, such as the Day One, Day Two, and Day Four creation accounts. Unfortunately, far too many people instead use the ANE cosmologies and creation myths to make those inferences and conclusions. That is, the beliefs and customs of the cultures that the ancient Hebrews were immersed in are used to interpret the biblical cosmology.

Even the ancient Hebrews sometimes were guilty of interpreting the Bible in terms of the beliefs of the cultures surrounding them. The Septuagint was a translation of the Hebrew Old Testament into Greek completed in Alexandria, Egypt, probably in the third century BC. Founded by Alexander the Great, Alexandria quickly became a thriving international city of trade, culture, and learning. The latter is evidenced by Alexandria's legendary library. While the Jews still living in Israel continued to read the Scriptures in Hebrew, many Jews of the Diaspora did not. Many Jews, such as the ones in North Africa, became Hellenized, and Hebrew became a lost language to them. Through Alexander's conquest and its aftermath, Greek became the international language and would remain so until eventually replaced by Latin in the latter years of the Roman Empire (at least in the West). With a desire among the Jews to read the Scriptures in their common language, a group of Jewish scholars in Alexandria reportedly translated the Septuagint into Greek. The word *Septuagint* derives from the word for 70, reflecting at least two legends about the Septuagint. One legend is that 72 scholars worked on the translation; another is that it took them 70 days to complete the translation. Most scholars today doubt the veracity of these legends, but the truth of the legend's details are immaterial to our discussion here.

In Genesis 1, the Hebrew word *rāqîaʿ* appears nine times (it appears only eight more times elsewhere in the Old Testament). The *rāqîaʿ* is that which God made on Day Two to separate waters above from waters below. The King James Version translated this word as "firmament," though more modern translations usually render it "expanse." The Septuagint translators chose the Greek word *stereoma* to translate *rāqîaʻ*. The word *stereoma* refers to a crystalline structure, something hard and transparent. In ancient Greek cosmology, the earth was surrounded by a hard, transparent sphere to which the stars were affixed. It appears that the translators of the Septuagint chose to equate the *rāqîaʻ* with the *stereoma* in order to incorporate into the Scriptures the cosmology of the ancient Greeks. In translating the Bible into Latin (the *Vulgate*), Jerome followed suit with the Septuagint and used the word *firmamentum*, which the King James Version transliterated into English. Jerome had good knowledge of the Hebrew language, so his choice in reflecting the Septuagint's translation probably was the result of his not knowing the meaning of the rare word *rāqîaʻ*.

What was this *rāqîaʻ* that God made on Day Two? Many people today would argue that the *rāqîaʻ* was the physical dome or solid vault of the sky that is part of ANE cosmologies. In many respects, this dome is similar to the *stereoma*. But this is to interpret the Old Testament in light of the ANE cultural beliefs. The Septuagint's treatment of *rāqîaʻ* might seem to suggest this, as well as medieval Jewish writings about cosmology. However, since these sources are much later than the writing of Genesis (by more than a millennium), all that this is evidence of is how the Jews *eventually came to view* the Genesis creation account. The translators of the Septuagint were Hellenized and hence their translation reflected Greek culture, not the ANE culture. And the medieval sources reflect their times, with an embrace of Aristotelean thinking, which in turn was a product of the ancient Greek worldview. It is one thing for unbelievers to view Genesis as a product of its times, but it is an entirely different matter for true believers in Christ to view Genesis this way. Ultimately, to view Genesis as merely a product of the surrounding ANE culture is to undermine the inspiration and authority of Scripture.

Supporters of such an interpretation of the Genesis creation account argue that the ancient Hebrews could not have understood the true cosmology (as we understand it today), so God revealed theological truth in a manner that they could understand, even if that implied cosmology was wrong. The faulty cosmological account is merely a vehicle for the communication, supposedly, of higher truth. Of course, this assumes that our cosmology of today is correct. But this also overlooks the fact that the Genesis creation account gives very little detail of cosmology, and instead gives a very brief description of cosmogony, the creation of the world. The creation account merely states that God made the *rāqîaʻ* on Day Two without revealing exactly what this *rāqîaʻ* is.

Even with all that we supposedly know today, there probably is more debate now than there ever has been as to the identity of the *rāqîaʿ*. That is the beauty of this aspect of the creation account: it adequately describes what God did on Day Two (and Day Four) without endorsing any particular cosmology. We can be certain that God made the *rāqîaʿ* on Day Two, though even today we may not understand any better than the ancient Hebrews what the *rāqîaʿ* is. The ancient Hebrews may have come to *interpret* what the *rāqîaʿ* was in terms of the ANE or Greek cosmologies, just as we today tend to interpret what the *rāqîaʿ* is in terms of our modern cosmologies (such as identifying it with space, or the earth's atmosphere, or both). But this is a far cry from concluding that the *original text* necessarily reflected any particular cosmology of the ANE. The assumption that the Genesis creation account reflects the faulty cosmology of the ANE undermines scriptural authority.

With these caveats about cultural influences on biblical interpretation in mind, what is the most likely identity of the *rāqîaʿ*? Clearly, the author of this book rejects the notion that it is some hard dome of the sky. At the risk of being guilty of interpreting the *rāqîaʿ* in terms of what we know (or think that we know) about modern astronomy and cosmology, let us work through this. As previously mentioned, God made the *rāqîaʿ* to separate waters below from waters above on Day Two of the Creation Week (Genesis 1:6–7). The word *rāqîaʿ* appears four times in the Day Two account. A second clue is provided by that statement immediately following and concluding the Day Two account that God called the *rāqîaʿ* heaven (*šāmayim*) (Genesis 1:8). The Hebrew word *šāmayim* appears in the Old Testament far more often than *rāqîaʿ* does, more than 400 times. The word *šāmayim* is plural; it lacks a known singular form in Hebrew. This introduces a problem for translation into English, because our equivalent word, heaven, has distinct singular and plural forms. The word *šāmayim* usually appears as an object in a sentence rather than the subject, so the verb used typically is of no help in determining whether the English translation ought to be singular or plural. That generally is left as the choice of the translator, which explains why various translations may differ on whether *šāmayim* is translated as plural or singular.[6] Unfortunately, this choice often reflects the cosmology assumed by the translator.

---

[6] For instance, in the first appearance of *šāmayim* in the Bible (Genesis 1:1), the King James Version translated it as "heaven," while the English Standard Version translated it as "heavens."

The word *rāqîaʿ* does not appear again until the Day Four account (Genesis 1:14–19), but *šāmayim* occurs right away in the next verse (Genesis 1:9) in the Day Three account, which states that God gathered the waters under the heaven(s) into one place to form the seas. Since Genesis 1:9 immediately follows Genesis 1:8, we may argue on the basis of context that the heaven(s) of these two verses must be the same thing. This is further reinforced by the fact that the entity made on Day Two was to separate the waters above it from the waters below it, and the Day Three account clearly concentrated on the waters below. The next time that the word *šāmayim* appears is in the Day Four account, the creation of astronomical bodies, where it appears three times (Genesis 1:14, 15, 17). All three times it appears in conjunction with *rāqîaʿ*. This combination reads "firmament of heaven" in the King James Version, though it appears as "expanse of the heavens" in the English Standard Version. This repetition may be significant. This construction seems to emphasize the connection of *rāqîaʿ* with *šāmayim*, with the point being that the thing in which God placed the astronomical bodies on Day Four is the same thing that God made on Day Two.

The word *rāqîaʿ* occurs one more time in the first chapter of Genesis (verse 20), during the Day Five account. This is in the context of the creation of birds, which the King James Version describes as flying "above the earth in the open firmament of heaven." The English Standard Version describes the birds as flying "above the earth across the expanse of the heavens." This language is different from how the firmament of heaven is presented in the Day Four account. First, it is described as being above the earth, which does not appear in the Day Four account. Furthermore, there is additional language ("in the open" in the King James Version, and "across" in the English Standard Version) modifying the firmament, or expanse, of heaven. The Hebrew expressed here is difficult to translate into English. It may indicate that the birds were to fly across the near face, or surface of the *rāqîaʿ*. If that is the case, then the *rāqîaʿ* encompasses the sky right above our heads, but much more beyond that. That is, the sky right above our heads is the near surface of the *rāqîaʿ*, but the *rāqîaʿ* extends much farther.

If this is the case, then what emerges is the understanding that the *rāqîaʿ* corresponds to what we today would call the lower parts of the earth's atmosphere and everything beyond—including outer space. Of course, this is our version of cosmology today; the ancient Hebrews would not have made such a distinction between the atmosphere and outer space. Perhaps God did not get more specific here, because if He did, it would have required endorsing some particular version of cosmology. If God had endorsed a common ancient view, it would be dismissed today as being wrong and hence undermining the Bible's veracity and authority. If God had endorsed modern cosmology, the ancient Hebrews would not have comprehended it, which likewise could have laid Scripture open to the accusation then that it was wrong. And who is to say that our modern cosmology is the correct one? There is no definite line of demarcation between the earth's atmosphere and outer space. We arbitrarily define space to begin at some specific altitude. However, there are at least two standards of defining the altitude where space begins. One standard is 50 mi (80 km), while the other is 100 km (62 mi). The fact that there are two distinct standards underscores the fact that the standard is somewhat arbitrary. One easily could change that definition to be a bit higher or lower. Hence, one could consider both the atmosphere and outer space part of the same thing, albeit with a gradient of how much air is present. Indeed, the biblical terminology surrounding *rāqîaʿ* is consistent with its being what we call outer space along with much, if not all, of the atmosphere.

Over the past half century, it has been common for biblical creationists to view the *rāqîaʿ* as the earth's atmosphere. This largely was motivated by support for the *canopy theory*. The canopy theory proposes that the earth originally was surrounded by water above the atmosphere (the waters above) that collapsed to provide one of the two sources of water for the Flood explicitly mentioned in the Flood account (the windows of heaven, as opposed to the fountains of the great deep, in Genesis 7:11). Support for the canopy theory has waned considerably in recent years, which should have been accompanied by the erosion of support for the idea that the *rāqîaʿ* is the earth's atmosphere—but that concept has endured. Integral to the equation that the *rāqîaʿ* is the earth's atmosphere is the belief that the first mention of the heavens in Genesis 1:1 refers to the creation of outer space. However, as previously discussed, this separation between the atmosphere and space is modern, and so may not coincide with what is meant in Genesis 1. Certainly, the ancient Hebrews would not have understood Genesis 1 this way.

Also note that Genesis 1:8 equates the *rāqîaʿ* with heaven. If Genesis 1:1 is a declaration of the specific act of the creation of heaven at the beginning of Day One, then it appears that God created heaven again on Day Two. One way out of this dilemma is to suggest that two different heavens were created on these days, outer space on Day One, and the earth's atmosphere on Day Two. As already pointed out, part of the motivation of this understanding was to support the canopy theory. Absent the canopy theory, is there a good reason to make this distinction? Further, is this interpretation supportable by the text? Would the ancient Hebrews have made this distinction? We know that later cultures, such as the Greeks, developed a concept of three heavens (the close environment of birds and clouds, the astronomical realm, and the abode of God), but there is no evidence that the Hebrews had this concept at the time of the writing of Genesis.[7]

A better resolution to this dilemma is to recognize that Genesis 1:1 serves as an introductory encapsulation, a summary of the creation account, with details to follow. In this manner, the phrase "the heaven and the earth" is a merism, an expression of totality by contrasting parts. There are several examples of merisms in English, such as "high and low," "young and old," and even "heaven and earth" (the Bible contains many merisms: Ecclesiastes 3:1–10 contains 14 merisms, while Romans 8:38–39 has several merisms). Thus, Genesis 1:1 is a bold and powerful statement that God made *everything*. This is reinforced by John 1:1–3, which reads,

> In the beginning was the Word, and the Word was with God, and the Word was God. He was in the beginning with God. All things were made through him, and without him was not any thing made that was made.

The Apostle John went on to identify this Word as Jesus Christ. Jesus is God, and as such, He is the Creator. But John explicitly states two different ways (both positively and negatively) that Jesus created everything. The creation account of Genesis 1 could have stopped with the first verse, but God wanted us to know some of the details of creation, such as the fact that He created the world in six days and rested on the seventh day. He wanted us to know that

---

[7] The only mention in the Bible of this three-tiered cosmology is in the New Testament, in 2 Corinthians 2:21. Since cosmology is not the point of this passage, it is not clear that this amounts to a biblical endorsement of this cosmology. Even if it is, this is not a problem, because this three-tiered cosmology conforms to our modern understanding.

He created man on the sixth day, along with land animals. He wanted us to know that He made the stars on the fourth day and that He made the *rāqîaʿ* on the second day. Furthermore, God wanted us to know some detail as to how he made man and woman, and what marriage is, lest there be any confusion about that, so He gave us more detailed information of the creation of man in Genesis 2.

We also ought not to be confused about how old the creation is. Some people may claim that if Genesis 1:1 is an introductory encapsulation, then it leads to belief that the creation is billions of years old. However, that would detach the introductory statement from the account that it heads. It is the details of the creation account that give us specific information about the creation. One cannot make an alleged inference about the broad introductory statement and then use that to reinterpret specific details from the following account. Rather, it is the specific details that allow us properly to understand the introductory encapsulation. There are strong textual reasons why the six days of creation were normal days, not long periods of time. Furthermore, the historical narrative of Genesis flows from creation through the death of Joseph in Egypt. The genealogies, along with specified lifespans, in Genesis 5 and 11 strongly indicate that the creation was just thousands of years ago, not billions of years ago. Therefore, viewing Genesis 1:1 as an introductory encapsulation does not lead to belief in billions of years. However, it does allow us properly to understand that God made the heavens (what we today would call the atmosphere *and* outer space) on Day Two.

So how ought we to view the Bible in relation to the creation mythologies and cosmologies of the ANE? The proper view is that, through the inspiration of God, the Bible records the true account of the creation of the world and its early history. The pagan sources do not. At best, those other sources relay garbled versions of the truth. For instance, we saw earlier that many pagan cosmogonies and theogonies prominently included water. Water also figures prominently in the creation account of Genesis 1. We have seen that on Day Two God separated water above and below by the *rāqîaʿ*, and that on Day Three God gathered together the waters below the *rāqîaʿ* to form the seas. However, the first mention of water is the deep (in Hebrew, *tᵉhôm*) in the first half of Genesis 1:2, referring to the state that God initially created the earth in on Day One. The Hebrew word *tᵉhôm* is equivalent to the Greek word *abussos*, from which we get the word *abyss*, referring to very deep water. The first time that the Hebrew word for water, *māyim*, appears is in the second half of Genesis 1:2, which states that the Spirit of God moved, or hovered, over the surface of the waters. From this we glean that God created the earth on Day One entirely out of water or totally covered by water, and that it remained so until Day Three. The word for water appears five times in the Day Two account, and twice on Day Three. On Day Five, God turned His attention back to the water to create sea creatures. The word for water appears three times in the Day Five account. The importance of water in the creation is repeated in 2 Peter 3:5–6, where the word *water* (*hudatŏs* in Greek) appears three times:

> For they deliberately overlook this fact, that the heavens existed long ago, and the earth was formed out of water and through water by the word of God, and that by means of these the world that then existed was deluged with water and perished.

This passage again emphasizes the importance of water in the creation. Even more significantly, water played a key role in the destruction of the world by the Flood. Notably, the Apostle Peter went on to tie the past judgment of the world by water to the coming judgment of the world by fire. Given the importance of water in the true account of the creation of the world, it is not surprising that various mythologies about the creation of the world would include that element. The cultural memory of the Flood may have contributed to this prominent inclusion of water in those mythologies, if nothing else, by convolving creation and the Flood. Likewise, it is not surprising that as ancient cultures developed their cosmologies, they convolved them with what little true knowledge of creation they had retained. Therefore, instead of the Genesis creation account reflecting the ANE cultures, it is the ANE cultures that very poorly reflect the truth of creation as found in Genesis.

Does the Earth Move?

As we saw earlier, the conflict thesis has heavily influenced our cultural mythology. This mythology teaches that the church supposedly held to the notion of a flat earth and opposed the idea that the earth is spherical. Another part of this mythology is that the church actively opposed the concept of the earth orbiting the sun, opting instead for an earth-centered universe. Unlike the flat earth myth, there is some truth to this story, though the details are a bit overwrought. We ought to begin with a few definitions. The *geocentric theory* is the belief that the earth is the center of the universe. That term comes from two Greek words: *geo*, meaning earth, and *kentron*, meaning sharp point, such as the point used on a compass to draw a circle (we get the word *center* from this root). The *heliocentric theory* is the belief that the earth is just one of several planets orbiting the sun. It comes from the Greek words *helios*, meaning sun, and *kentron*. Notice that unlike the geocentric theory, where the earth is the center of everything, in the heliocentric theory, the sun merely is the center of the solar system.

Throughout history, the majority viewpoint was the geocentric theory, but this began to change four centuries ago. How did belief in the geocentric theory come about, and why did it persist so long? Our cultural mythology is that people in the past were superstitious and didn't know any better. However, when asked, most people today cannot give a single reason how we know that the heliocentric theory is true. They think that they know what is true, and they have faith that science somehow has proven it. But how is this different from the "ignorant" people of the past?

A common belief today is that people in the past thought that since humanity was the center of God's attention, we must be at the physical center of creation. This is anachronistic, because better reasons came first. Furthermore, being at the center was far from being in a favored location. In the Aristotelian worldview, things naturally moved downward. That is, things fell toward the earth's center, because that was the natural direction that things went. There is equivocation here, because this rule applied not only physically, but also morally. Man obviously was corrupt, but the ancient Greeks thought that the world (the earth) in which man lived was corrupted as well (this probably was a vestige of the truth of Genesis 3). This is in contrast with the heavenly realm, where the gods lived. Things were perfect in the heavenly realm, but the earthly realm was far from perfect. The center of the earth was as far from the heavenly realm as one could be, so the earth's center was the lowest place, both physically and morally. Thus, being on earth's surface, man was perched right above perdition. Far from being in a favored location, man was only one level above the least favored location. Therefore, the necessity of the geocentric cosmology is a pagan Greek concept, not Christian.

We shall discuss in greater depth the motions that we see in the sky shortly, but we can consider them at some level now. Everyone is familiar with the fact that the sun rises in the east and, after moving across the sky, sets in the west each day. There can be two causes of this motion: either the sun moves around the earth each day, or the earth turns on an axis each day. Which one of these is true? That is not immediately clear. Ancient cosmologies were a bit divided on this question. If the earth were flat, rotation of the earth around an axis does not make sense, so flat earth cosmologies generally had the sun move across the sky each day. Once people realized that the earth was spherical, it was possible to conceive that the earth might rotate each day. However, notions are difficult to change, so belief in a non-rotating earth often persisted even with a spherical earth. Less well known is that at night most stars, the planets (which appear as bright stars), and the moon also rise and set similar to how the sun does each day. Again, this could be explained either by the earth spinning or the sky spinning around the earth each day. Out of this observation developed the notion of the *stereoma*, which, as we saw before, was a hard crystalline sphere on which the sun, moon, and stars were affixed.

Ignoring the apparent rotation of the sky around the earth each day, other changes take place over longer time periods. From one night to the next, the moon moves about 13° eastward with respect to the stars, taking one month to complete one circuit. Both today and in ancient times, this motion was attributed to the moon orbiting the earth. The sun appears to move about 1° eastward through the stars each day, taking a year to return to its starting place. As with the daily motion of the sky, this can be interpreted two ways: either the sun orbits the earth with respect to the stars once per year, or the earth orbits the sun. More than daily motion, this annual motion of the sun through the stars is the major observation that the geocentric and heliocentric theories attempt to explain. Which theory is true? That question is more difficult than many people realize. Ancient philosophers and scientists developed many answers to this question. Some protested that if the earth moved, we ought to feel that motion. However, our ability to feel motion relies upon bumps and other irregularities in motion. We now recognize that motion can be so smooth that the sensation of motion is not possible. Another objection to the earth's motion was that the moon obviously orbits the earth each month, but if the earth moved, it would leave the moon behind.

However, the greatest objection to the heliocentric theory in the ancient world was an inability to observe *parallax*. Parallax is the apparent change in position that an object displays as we change our vantage point. You can demonstrate this by viewing a thumb held up at arm's length first with one eye and then the other. Your thumb will appear to shift left and right. In a similar manner, as the earth orbits the sun each year, stars appear to shift back and forth due to our changing position. Ancient astronomers watched for this parallax effect, but they never saw it. Being good scientists, they rejected the heliocentric theory in favor of the geocentric theory. That is, ancient scientists actually considered the heliocentric theory (rather than rejecting it on philosophical grounds) and discarded it because of the failure of prediction that the theory made. Lest you worry that you've been lied to about the earth's motion around the sun your entire life, the ancients did not see parallax because the stars are too far away. The amount of parallax decreases as the distance to stars increases. Stars are so far away that their parallax is extremely tiny, requiring a telescope. The first parallax measurement was not until the 1830s.

There were a few people in the ancient world who believed in the heliocentric theory anyway. The first person that we know of who taught the heliocentric theory was Aristarchus of Samos in the third century BC. Aristarchus based his heliocentric belief upon his estimates of the relative sizes of the earth, moon, and sun. He determined that the moon's diameter was one-third that of the earth, while the sun's diameter was seven times that of the earth.[8] It made more sense to Aristarchus that the smaller bodies should orbit the larger bodies. Since the sun was larger than the earth, the earth must orbit the sun. Aristarchus even got the ordering of the planets from the sun correct. He based this upon how fast the planets appeared to move, with the faster moving planets closest to the sun. Aristarchus, though, was not the first person to suggest that the earth moved, because a century earlier, Philolaus, a student of Pythagoras, taught that the earth moved. However, in Philolaus' system, the earth orbited a central fire, not the sun. How did Aristarchus explain the lack of observed parallax? He explained it the same way that we do today—stars are so far away that their parallax is too small to be seen by the eye. This was not a new concept either, because even earlier, in the fifth century BC, Anaxagoras had suggested that the stars were very distant objects similar to the sun. Anaxagoras argued that the stars appear so much fainter than the sun

---

[8] In reality, the moon is one-quarter the size of the earth, and the sun is 109 times larger than the earth.

because of their tremendous distance. For some time, historians had thought that heliocentric view as taught by Aristarchus was a small minority viewpoint in the ancient world. However, in recent years, some historians have begun to think that the heliocentric theory may have been more popular than previously believed. At any rate, it probably remained in the minority, and by the second century AD belief in the heliocentric theory had disappeared. Part of the reason for the dominance of the geocentric theory undoubtedly was the tremendous influence of Aristotle, who predated Aristarchus by nearly a century. Aristotle clearly favored the geocentric theory, again upon the basis of lack of observed parallax.

Probably what cemented the dominance of the geocentric theory was the cosmology developed by Claudius Ptolemy in the second century AD. Ptolemy attempted to explain the motion of the planets, the one motion of heavenly bodies that we have not yet described. Due to the earth's rotation, the entire sky seems to spin around us once each day and night. The moon orbits the earth through the stars once a month. The sun appears to move through the stars once per year, causing seasonal changes in the stars we see at night. All of these motions are very regular. However, the five naked-eye planets, Mercury, Venus, Mars, Jupiter, and Saturn, appear as bright stars that move through the stars in a less regular manner. The planets usually move eastward among the stars, the same direction that the sun and moon move. However, from time to time, the planets reverse direction and travel westward with respect to the stars, before once again reversing direction and traveling eastward again. Astronomers call the normal eastward motion of planets *prograde*, or direct, motion, and they call the westward movement *retrograde*, or indirect, motion. Both prograde and retrograde motion generally are along the ecliptic, the projection of the earth's orbital plane on the sky delineated by the sun's apparent annual motion. However, the motions of the planets are inclined slightly to the ecliptic, intersecting in two points called *nodes*. The moon's motion similarly is tilted to the ecliptic a few degrees.

How do these motions take place? Consider a planet farther from the sun than the earth, such as Mars. With a larger orbit, Mars has greater distance to travel to complete one orbit. Furthermore, being farther from the sun, the attraction of the sun's gravity is less than that on the earth, so Mars moves more slowly than the earth does. Consequently, it takes longer for Mars to complete one orbit around the sun, about two years instead of one year for the earth. Figure 1.5 shows what happens as the earth overtakes and then passes Mars. The numbered lines illustrate the direction that one must look to see Mars from various points on the earth's orbit. Those lines are projected onto background stars. Notice that at first the projection of Mars onto the stars increases upward on the diagram. This illustrates prograde motion (eastward). However, as the earth overtakes Mars, the projection of Mars' position appears to reverse, going retrograde (westward). Once the earth has passed Mars sufficiently, the projections again reverse direction and Mars appears to move prograde (eastward) again. A similar thing happens when Mercury and Venus, planets with orbits smaller than earth's orbit, pass the earth. The slight bobbing up and down of the planets (and the moon) with respect to the ecliptic is due to the inclination of each planet's (and the moon's) orbit with respect to the ecliptic.

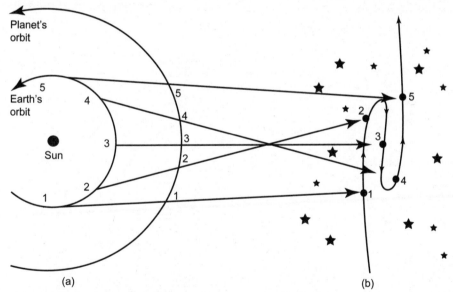

(a)   (b)

Fig. 1.5. Earth passing Mars in orbit relative to star field (Danny R. Faulkner, *The New Astronomy Book* [Green Forest, Arkansas: Master Books, 2004], 11).

While the behavior of the planets' motion is relatively easy to explain in terms of the heliocentric theory, it is much more difficult to explain with the geocentric theory. One could suppose that the planets simply moved in one direction for a while as they orbited the earth, but then reversed direction for a while before returning to normal, prograde motion. However, there is no predictive power in that proposal. Furthermore, as previously mentioned, the ancients thought that the circle was the perfect shape, and since the heavenly bodies were considered to be perfect, they must follow perfect, circular motion. This concept of perfection extended to the insistence that the motion along circles be uniform. But motion cannot be uniform if it reverses direction. Figure 1.6 illustrates Ptolemy's solution to this problem. Ptolemy had each planet move uniformly on a circle called an epicycle. The epicycle in turn uniformly moved on a larger circle called a deferent. Normally, a planet's motion along its epicycle combined with the epicycle's motion along the deferent to produce prograde motion. However, there is a portion of an epicycle that is closest to earth. When a planet passed through this part of its epicycle, its motion appeared from earth to oppose the motion of the epicycle on the deferent. The effect is that the planet appears to move backward for a while. With adjustment of the sizes of the epicycle and deferent, along with the two speeds of motion, Ptolemy produced a good fit to observations.

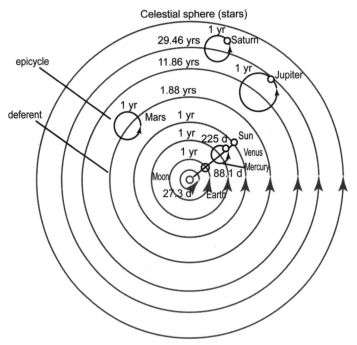

Figure 1.6. Illustration of Ptolemy's epicycles (Danny R. Faulkner, *Universe by Design* (Green Forest, Arkansas: Master Books, 2004), 12.

As it was, though, Ptolemy's theory required additional adjustments. Ptolemy explained the planets' small bobbing up and down from the ecliptic by a second, smaller epicycle perpendicular to the first, larger epicycle required for each planet. An epicycle of this type also was required for the moon's motion. To further improve the fit with planetary data, Ptolemy displaced the earth slightly from the center of each deferent. This was necessitated by the fact that planetary orbits are not circular, but instead are ellipses, with the sun at one focus (not at the center) of each ellipse. However, the orbits of the planets are not so elliptical that they cannot be approximated by circles with the earth off-center. The second refinement was that each epicycle did not move at a uniform rate with respect to the center of its deferent, but instead it moved at a uniform rate with respect to the *equant*. The equant of each deferent was the point collinear with the earth and the center of the deferent, but opposite and the same distance from the center as the earth. This was required by the fact that the planets do not move at a uniform rate along their orbits around the sun, but rather they move at rates so that the line between each planet and the sun sweeps out equal areas in equal time intervals. These latter two adjustments, displacing the earth from the center and the redefined manner in which the planets uniformly moved, were departures from the basic assumptions of the geocentric model, but apparently not many people objected to that. Perhaps they were impressed by the fact that the Ptolemaic model worked pretty well to predict planetary positions, where no theory prior to the Ptolemaic model had. At any rate, 15 centuries later, Johannes Kepler would reformulate these two major adjustments within the heliocentric theory in the form of the first two of his three laws of planetary motion.

The next series of developments is especially important in the history of astronomy. It appears that Ptolemy did not intend that his model be taken literally. Rather, he merely was describing the motion of the planets, and he used the best mathematics available to him, geometry. Today we use algebraic expressions to describe planetary motion, but no one thinks that the algebraic expressions themselves are reality. However, people eventually came to think of the circles of the Ptolemaic model as reality. That is, people believed that the planets actually moved along these circles.

A few centuries after Ptolemy, the Roman Empire collapsed, leaving chaos and turmoil in its wake. There was serious erosion of commerce, technology, and learning that led to the Middle Ages. Late in the Roman Empire, Christianity had shifted from being a persecuted sect, to a tolerated religion, to *the* state religion. This move toward acceptance and then compulsion inevitably led to the inclusion of many unregenerate people within what was recognized as the church. In the West, this corrupted church began to centralize in Rome, the original capital of the Roman Empire. As people longed for the security that Rome had provided in the past, people increasingly came to view the Roman Church as its leader. Thus, the Roman Catholic Church came to dominate not only theology, but also academics and political power. A millennium later Europe emerged into the Renaissance. By the so-called Enlightenment of the seventeenth century, it became fashionable to call the Middle Ages the Dark Ages and to blame Christianity for the erosion of learning during the Middle Ages. Ultimately, this led to the previously-discussed conflict thesis. The Middle Ages were not that dark, as many historians now acknowledge. Furthermore, the great losses of the Middle Ages occurred early, before the Roman Catholic Church came to dominate, so the Roman Catholic Church hardly could be blamed for the problem. Instead, the dominance of the Roman Catholic Church came about in response to crises that Europe faced.

In the waning days of Rome, Augustine was bishop of Hippo, in North Africa. Augustine's writings were extremely influential in the development of western philosophy and Christian theology, and are well regarded today by Roman Catholics and some Protestants. Augustine was heavily influenced by the writings of Plato, and he incorporated Platonism or, more properly, Neoplatonism into his philosophy and theology. It is a shame that Augustine did not attempt to build a truly biblical worldview from the Scriptures,[9] but instead chose to build a pseudo-biblical worldview on a foundation that included pagan ideas from the ancient Greeks. His approach allowed for, among other things, the belief that men using their rational minds could arrive at spiritual truth (which undermined a biblical perspective on the noetic effects of sin and the need for biblical revelation). One consequence of Augustine's teachings was the de-emphasis of Aristotle in the West. Unlike Plato, who was more of a philosopher, Aristotle was both a philosopher and

---

[9] For a discussion of the importance of Scripture in the development of a biblical worldview, see Danny R. Faulkner, *The Created Cosmos: What the Bible Reveals about Astronomy* (Green Forest, Arkansas: Master Books, 2016), 323–333.

a scientist. Through Augustine's influence, the works of Aristotle and other ancient scientists virtually disappeared in the West. This probably led to the common belief today that in the Middle Ages people did not conduct science as we know it to explore the world around them, but instead philosophized about the world. One of those scientists whose work was lost in the West was Ptolemy.

However, the works of Aristotle, Ptolemy, and others remained in circulation in what had been the eastern part of the Roman Empire. In the wake of the Islamic conquest of the seventh century, Muslims came into contact with these works, which they translated into Arabic. Of particular interest to the Muslims was the work of Ptolemy. The Muslims were so impressed with the Ptolemaic model's ability to predict planetary positions that they began to call Ptolemy's work "the Greatest," or, in Arabic, *Almagest*. This is the title by which Ptolemy's work is known today. By the twelfth century, translations of Aristotle, the *Almagest*, and other ancient Greek works into Latin and even modern European languages began to appear in the West. In the thirteenth century, Thomas Aquinas did for Aristotle what Augustine had done for Plato eight centuries earlier. Soon Aristotelian philosophy held sway in Western thought, and this undoubtedly provided fertile ground for renewed interest in the pursuit of science in the West. Augustine had laid the groundwork for accepting man's ideas as ultimate truth, but Aquinas more fully developed it. Aristotelianism became the filter through which Scripture was viewed. Since Aristotle had taught the geocentric theory, certain Bible passages, such as the account of the long day at the Battle of Gibeon in Joshua 10, were interpreted in terms of the geocentric theory. That is, it became *doctrine* that the Bible allegedly taught the geocentric theory. The Bible does no such thing, but this nevertheless eventually became the general belief. This is an excellent example of *eisegesis*, reading a foreign meaning into the Bible. By contrast, our approach ought to be to strive for sound *exegesis*, that is, the practice of drawing the author's intended meaning out of the text, and so determining what Scripture actually teaches. Only then can we properly test our ideas.

This change in attitude did not happen immediately, but it was fully in place by the early seventeenth century. At that time, the Roman Catholic Church certainly felt that it was under assault. The Protestant Reformation had exploded nearly a century before, and it had led to much loss of influence in parts of northern and western Europe. At the same time, Islam was on the prowl gobbling up portions of Southeastern Europe. The Roman Catholic Church did not have the stomach for more attacks upon its authority. It was at this time that Galileo Galilei began to present a problem in northern Italy. Galileo was a brilliant mathematician and scientist. As a young man, Galileo had read the book by Nicolaus Copernicus, written a few decades earlier, in which Copernicus espoused the heliocentric theory. Copernicus often is credited with inventing the heliocentric theory, but, as we saw earlier, this is not true. Rather, Copernicus resurrected, developed, and popularized the heliocentric theory. The primary argument for the heliocentric theory was that it was much simpler than the geocentric theory. Since the time of Ptolemy, people had found that the Ptolemaic model required a number of additional adjustments, such as many more epicycles, to explain planetary motion adequately. Surprisingly, though Copernicus' book challenged the official teaching of the Roman Catholic Church and was widely read, it had received relatively little condemnation.

Even Galileo garnered little attention at first. While Galileo already believed the heliocentric theory when he began to use a telescope, his telescope provided evidence that the Ptolemaic model was wrong. Contrary to popular belief, Galileo did not invent the telescope, but he does appear to be the first person to put the telescope to use to study astronomical bodies. With his telescope, Galileo saw that the planet Venus went through a complete set of phases similar to the moon. Unlike the moon, which can be anywhere along the ecliptic, Venus is never more than 47° from the sun. For Venus to exhibit a full range of phases, it must orbit the sun. Hence, this simple observation disproved the Ptolemaic model, opening the door to the heliocentric theory. One argument that Aristotle had advanced against the heliocentric theory was that if the earth moved, the moon would be left behind. However, through his telescope, Galileo saw that four satellites, or moons, orbited Jupiter. It was clear that Jupiter moved, even in the Ptolemaic model, yet its satellites were not left behind as it moved. Therefore, Aristotle's objection to the earth moving was not correct. Galileo found other challenges to Aristotelian thought. For instance, Galileo saw craters on the lunar surface and spots on the solar surface. Both the sun and the moon, being heavenly bodies, were deemed perfect, and perfect bodies could not have craters and spots.

These observations emboldened Galileo, and as he more forcefully taught the heliocentric theory, Galileo began to come under attack. However, those attacks came not from theologians but from other scientists, who complained that Galileo was contradicting established science. In response, in 1615, the Roman Catholic Church ruled that the heliocentric theory could be discussed as a theory, but not presented as fact, since it contradicted the Ptolemaic model, which the Roman Catholic Church had endorsed. Seventeen years later, in 1632, Galileo published a popular book that not only espoused the heliocentric model, but also ridiculed the Pope, as well as many of Galileo's critics. Highly offended, these leaders called for Galileo's discipline, which led to a trial. Galileo was accused of several charges. Some of these charges, such as insubordination, he clearly was guilty of. At the conclusion of his trial, Galileo was found guilty, required to recant, and was sentenced to house arrest. Now elderly, Galileo spent the rest of his life under house arrest. Contrary to popular belief, Galileo was neither charged with nor found guilty of heresy. If he had, he likely would have suffered far worse punishment, even unto death—other people at the time certainly did. The issue was not even theological in nature, but instead was scientific; and much of the argument against the heliocentric theory in the trial came not from Scripture, but from Aristotle and Ptolemy.

Galileo's opponents were fighting a losing battle. Already, support for the Ptolemaic model was rapidly eroding. In 1600, nearly everyone believed the Ptolemaic model. By 1700, no one did. Why was there such a rapid shift? The Ptolemaic model was very complicated. Over the years, there were slight discrepancies between the predictions of the model and the observed locations of the planets. The fix to this problem was to add additional epicycles. While this modified system correctly predicted planetary positions, it was unwieldy. Some versions required a hundred epicycles. It is said that in the thirteenth century, when Alfonso X of Castile was schooled in the Ptolemaic model, he reportedly commented that, "If the Lord Almighty had consulted me before embarking on creation thus, I should have recommended something simpler." A general principle of reasoning is that the simplest explanation probably is the correct one (this often is called *Occam's razor*, for William of Occam of the fourteenth century). By the seventeenth century, many people began to realize that the heliocentric model's explanation for retrograde motion was much simpler than that of the Ptolemaic model. This paved the way for getting out of the rut of Aristotelian thinking, such as constraining heavenly bodies to move uniformly in circles. This soon led to Newton's formulation of gravity and the development of physics. Galileo provided the first direct evidence (the phases of Venus) that the Ptolemaic model was wrong, but it was not until the eighteenth century that the first direct evidence (the aberration of starlight) that the heliocentric theory was true was seen.

CHAPTER 2

The Expanse of the Heavens: Observing the Sky Above

What Does the Sky Look Like?

The sky has the appearance of an inverted bowl. This is why the domed ceiling of a planetarium provides a realistic representation of the heavens. As discussed in Chapter 1, the earth's rotation causes celestial bodies to move around us each day. Therefore, rather than being a bowl, the sky appears as half a sphere, quite fittingly called the *celestial sphere* (Figure 2.1). This is what the *stereoma* in

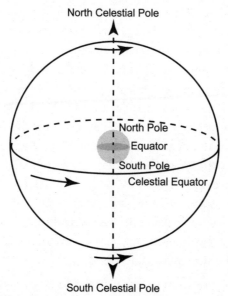

Figure 2.1. Concept art of celestial sphere with celestial north pole.

ancient Greek cosmology was. However, unlike the *stereoma*, which was an actual, hard, transparent sphere, the celestial sphere is an *imaginary* sphere. The celestial sphere is a useful theory. A theory, as such, is just a detailed explanation of phenomena. Theories often are unfairly pitted against facts. For instance, someone may dismiss an idea by saying, "That's just a theory," as if to imply that the idea is not true. We use many theories in science that we think are true. One example is Newton's theory of gravity.[1] However, other theories, such as the celestial sphere, we know are not true, but we use them because we find them useful in explaining things.

The earth's rotation causes the celestial sphere to spin around us each day. There are two spots on the celestial sphere around which the whole celestial sphere appears to turn. These two spots are the intersections of the earth's rotation axis with the celestial sphere. One of these spots is the north celestial pole, and the other is the south celestial pole. On the Northern Hemisphere of the earth, one can see the north celestial pole, but not the south celestial pole. The reverse is true in the Southern Hemisphere. The north celestial pole makes an angle with respect to the northern horizon that is equal to a person's latitude. As the earth spins, stars appear to revolve counterclockwise around the north celestial pole. Polaris, or the North Star, is within a degree of the north celestial pole, so to the naked eye, the North Star does not appear to move. Rather, the North Star appears fixed above the northern horizon, at an angle equal to one's latitude. We have given this name to the North Star, because it indicates which direction is north. However, the North Star has not always been so favorably placed, nor will it always be. A subtle effect called precession of the equinoxes causes all of stars slowly to change position in the sky. In a thousand years, Polaris will not be so close to the north celestial pole. By the way, there is no corresponding bright star near the south celestial pole.

Stars near the north celestial pole neither rise nor set in the night sky, but are continually above the horizon as the celestial sphere spins. We say that these stars are *circumpolar*, meaning "around the pole." All stars within an angle from the north celestial pole equal to one's latitude are circumpolar. Except at very high latitudes on earth, most stars are not circumpolar. Instead, most stars rise and set, as do the sun, moon, and planets.

---

[1] Even Newton's law of gravity, while generally true, has been superseded by Einstein's theory of general relativity. However, the difference between the two theories only arises in certain circumstances. If Newton's theory is used within its bounds, physicists consider it to be true.

Halfway between the north and south celestial poles is a circle that we call the *celestial equator*. As with the earth's equator, the celestial equator divides the celestial sphere into two hemispheres—the north and south celestial hemispheres. The celestial equator is an example of a *great circle*. A great circle is the intersection of a sphere and a plane that passes through the center of the sphere.[2] Another example of a great circle on the celestial sphere is the ecliptic, the plane of the earth's orbit. The ecliptic is tilted with respect to the celestial equator by the same angle of tilt of the earth's axis, approximately 23.5°. Astronomers call this angle the *obliquity of the ecliptic*.

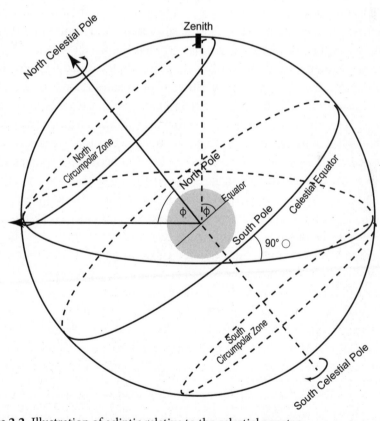

Figure 2.2. Illustration of ecliptic relative to the celestial equator.

---

[2]An example of a circle on a sphere that is not a great circle is a parallel of latitude. A parallel of latitude is the intersection of the earth's surface (a sphere) and a plane, but the plane does not pass through the earth's center.

All great circles on a sphere intersect in two places. The two points where the ecliptic and the celestial equator intersect are the *equinoxes*. The equinox where the sun is moving northward as it crosses the celestial equator is the *vernal equinox*, while the equinox where the sun is moving southward as it crosses the celestial equator is the *autumnal equinox*. From the names, you likely can infer that when the sun reaches the vernal equinox once per year, it ushers in spring, while when the sun is at the autumnal equinox it is the beginning of autumn. Halfway between the equinoxes on the ecliptic are the *solstices*. The solstices are the points on the ecliptic farthest north and south of the celestial equator. In the Northern Hemisphere, the *summer solstice* is the solstice that is north of the celestial equator, while the *winter solstice* is south of the celestial equator. We say that summer has begun when the sun reaches the summer solstice and that winter has begun when the sun has reached the winter solstice. Properly defined, the equinoxes and the solstices are points on the celestial sphere. But in common usage, the equinoxes and solstices refer to dates when the sun reaches these points on the celestial sphere. For instance, the sun reaches the summer solstice each year around June 22. We say that the summer solstice occurs on this date, though the summer solstice, as a point on the celestial sphere, always exists.

Of course, the seasons are reversed in the Southern Hemisphere, and so these terms as defined do not make sense in the Southern Hemisphere. The overwhelming majority (more than 90%) of the earth's population lives in the Northern Hemisphere. Furthermore, these terms were developed long ago, before there were any nations as we know them in the Southern Hemisphere. However, in deference to those who live in the Southern Hemisphere, it is becoming more common to call the equinoxes the March and September equinoxes rather than the vernal and autumnal equinoxes. Similarly, the solstices increasingly are called the June and December solstices rather than the summer and winter solstices.

We group stars together into patterns that we call *constellations*.[3] There are 88 constellations. More than half of them go back to ancient times. Although the stars move, they are so far away that even throughout human history, the

---

[3] I will not discuss the constellations in this book. In the companion book, (Danny R. Faulkner, *The Created Cosmos: What the Bible Reveals about Astronomy*. Green Forest, Arkansas: Master Books, 2016), I discuss a few of the constellations as they directly relate to the Bible, as well as to the gospel in the stars theory, a prevalent but misguided theory with regard to the origin of the constellations. If you wish to use star charts to study constellations, I suggest two books written by biblical creationists . One book is Jason Lisle's *The Stargazer's Guide to the Night Sky* (Green Forest, Arkansas: Master Books, 2012) and the other is Jay Ryan's *Signs and Seasons: Understanding the Elements of Classical Astronomy* (Cleveland, Ohio: Fourth Day Press, 2007).

basic shapes of the constellations have not exhibited discernable change. This is in stark contrast to the sun, moon, and the five naked-eye planets. These bright objects move quite noticeably with respect to the background of stars. However, these wandering objects cannot be found just anywhere on the celestial sphere. The sun's motion defines the ecliptic, so the sun always is on the ecliptic. Since the orbital planes of the moon and the other planets are inclined only slightly to the ecliptic, the moon and planets appear within a band a few degrees to either side of the ecliptic. There are 12 constellations along the ecliptic, and so the sun, moon, and planets normally are found only in these constellations, making these constellations special. You probably are familiar with these 12 constellations, because they are the constellations of the *zodiac*. The zodiac is the swath of the sky centered on the ecliptic where the sun, moon, and five naked-eye planets always are found. Why are the names of these 12 constellations, or signs, so familiar to us? To ancient pagans, the sun, moon, and planets were gods. Hence, the position of these gods could affect our lives and our destinies. Out of this belief came the ancient pagan cultic system of astrology. Many people today dabble in astrology, never realizing that it is an ancient pagan belief. Christians ought not to be involved with astrology. For more information about what is wrong with astrology, please see Chapter 5 of the companion book, *The Created Cosmos: What the Bible Reveals about Astronomy* (Faulkner 2016).

What Causes the Seasons?

Nearly everyone knows that the obliquity of the ecliptic is related to the change of the seasons. However, most people do not fully understand how the earth's tilt causes the seasons. This is the fault of poor teaching in many schools. Most people grasp that during summer we are tilted toward the sun and in winter we are tilted away from the sun. Unfortunately, far too many people think that it is the *changing distance* that accounts for the difference between summer and winter temperatures. This makes sense to most people, because the terms "toward" and "away" frequently refer to distance. Furthermore, most diagrams illustrating the seasons show the sun very close to the earth, so it easily appears as if our distance from the sun changes appreciably from summer to winter. However, these illustrations are horribly out of scale. The average distance between the earth and sun is *93,000,000 mi (15,0000,000 km)* while the diameter of the earth is a mere *8000 mi (13,000 km)*. Therefore, the point on the earth that is closest to the sun is less than $^1/_{100}$ of 1% closer to the sun than the point on the earth that is farthest from the sun. And the tilt alone would not change the distance that much. Is it reasonable that such a miniscule difference in distance could account for the difference in seasonal temperatures? Clearly, it is not. Furthermore, as we saw earlier, the planets' orbits are ellipses, not circles. Because of the earth's elliptical orbit, the earth is closest to the sun in early January and farthest from the sun in early July. Therefore, the earth is 3% closer (3,000,000 mi) to the sun in the Northern Hemisphere's winter than it is in the Northern Hemisphere's summer. If distance from the sun were the key factor in causing the seasons, then the entire earth would experience winter in July and summer in January.

If changing distance does not explain the seasons, then what does? There are two factors: the *area effect* and the *time effect*. When we are tilted away from the sun, the sun's rays, even at noon, strike the earth's surface at a much lower angle than when we are tilted toward the sun. Imagine a ray of sunlight as a column or cylinder. When a ray of sunlight strikes the ground at noon in summer, it makes a relatively small ellipse (Figure 2.3). However, that same ray will make a much larger ellipse at noon in winter, when it strikes the ground at a much lower angle. In either case, the ray of sunlight packs the same amount of energy (heat), but because during the winter the energy is spread out over a much larger area, it is far less concentrated than it is in summer. With far more heat delivered per unit of surface area in summer, it is warmer; but with far less heat per unit of area in winter, it is cooler. This is the area effect.

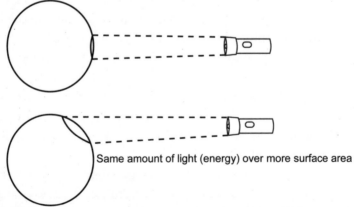

Same amount of light (energy) over more surface area

Figure 2.3. Illustration of area effect with a cylindrical beam of light.

What is the time effect? In the winter, the sun is far south of the celestial equator, so the sun rises well south of east and sets well south of west. With less distance on the celestial sphere to traverse from sunrise to sunset, winter days are short. However, in the summer the sun is well north of the celestial equator, and the sun rises north of east and sets north of west, resulting in a much longer path to travel throughout the day. Therefore, summer days are long. This explains why an earth that rotates at a constant rate can have different lengths of daylight and darkness. With the sun in the sky much more in the summer, there is ample time for the earth to absorb the sun's rays, making it warm. However, during the winter the days are short, providing relatively little time for absorbing the sun's energy. Conversely, winter nights are long, allowing for energy to escape into space, cooling the earth. But summer nights are short, which does not allow too much energy to escape the earth.

While maximum and minimum heating coincide with the solstices, average maximum and minimum temperatures occur about one month later. *Thermal inertia* causes this delay. Thermal inertia refers to the time it takes for objects to heat and cool. It takes time to heat the earth during the summer, and once the heat accumulates, it takes time to dissipate. This dissipation continues through late summer, autumn, and early winter until about a month past the solstice, whereupon there already is an increase in the daily input of heat so that the slow warming process begins anew. The warming continues through spring and early summer until about a month past the solstice, restarting the cycle.

Thermal inertia plays a key role in another interesting topic related to the seasons. As previously mentioned, earth's perihelion, the point of earth's closest approach to the sun, occurs in early January, while earth's aphelion, the point at which the earth is farthest from the sun, is in early July. In the Northern Hemisphere, this results in a little less heating during summer months and a little more heating in winter months than if the earth followed a circular orbit. Hence, winter and summer are milder than they would be if the earth's orbit were circular. However, this is true only for the Northern Hemisphere. Maximum heating occurs during southern summer and minimum heating during southern winter. Therefore, one might expect that the seasons in the Southern Hemisphere would be more extreme than they are in the Northern Hemisphere. However, they are not. How can this be? A look at a globe reveals that the surface area of the Northern Hemisphere is comprised of about 40%

land (and about 60% water), but that land accounts for only about 20% of the Southern Hemisphere. Water has a very high specific heat. That means that it takes much heat in order to warm water. Consequently, it takes a long time for water to heat or to cool. Therefore, water is effective in heating and cooling things. It also is why cities near large bodies of water, such as San Diego, have moderate climates. The large amount of water in the Southern Hemisphere acts as a thermal buffer to moderate temperatures both in summer and winter. The circumstances of this are ideal. If perihelion and aphelion were shifted by a few months from their current times, there would be greater extremes in Northern Hemisphere weather. This is evidence that the earth as it now exists is designed.

Some biblical creationists believe that before the Genesis Flood, the obliquity of the ecliptic was zero. The biblical basis for this supposedly is Genesis 8:22, which reads,

> While the earth remains, seedtime and harvest, cold and heat, summer and winter, day and night, shall not cease.

This verse is part of the Lord's discourse that immediately follows the Flood. Those who believe that the earth's tilt changed dramatically at the time of the Flood reason that what is mentioned in Genesis 8:22 was the introduction of something new, so that summer and winter had not existed prior to the Flood. If there were no summer and winter before the Flood, and if the seasons are caused by the obliquity of the ecliptic, then the obliquity of the ecliptic must have been zero before the Flood. But does this verse really indicate that there were no seasons prior to the Flood? Day and night are mentioned in this verse in the same context, yet we know that day and night existed before the Flood. Therefore, it does not follow that this verse indicates the institution of seasons as we know them. What does this verse mean? In Genesis 8:20–22, God states that He would not again destroy the earth because of man's wickedness as He had just done with the Flood. This sentiment is repeated in the Noahic Covenant of Genesis 9:1–17 (especially verses 11 and 15). Until the renewal of creation when there is a New Heaven and a New Earth (Isaiah 65:17; 66:22; 2 Peter 3:13; Revelation 21:1), things will go on uninterrupted by a worldwide cataclysm like the Flood. That is, the cycles of day and night, summer and winter, and seedtime and harvest will continue.

Associated with the belief that there was no tilt of the earth's axis before the Flood is a belief in some impulsive event associated with the Flood that caused the earth's tilt to change dramatically. A half century ago, the Australian astronomer and creationist George Dodwell thought that he had found evidence of a catastrophic change in the obliquity of the ecliptic. The most direct method for measuring the obliquity of the ecliptic involves the use of a *gnomon* (Figure 2.4). A gnomon is a vertical pole of known height that casts a shadow. The ratio of the height of the gnomon to the length of its shadow is equal to the tangent of the angle that the sun makes with the horizon (the altitude). If one measures the sun's altitude at noon on the winter and summer solstices, the difference is equal to twice the obliquity of the ecliptic. In the mid-fourth century BC, Pytheas of Massalia used this technique to measure the obliquity of the ecliptic near modern-day Marseilles, France. This is the earliest recorded measure, but others may have measured it earlier. By the time of Hipparchus two centuries later, this was the standard manner of measuring the obliquity of the ecliptic. By the Middle Ages, astronomers were aware that the obliquity of the ecliptic was not constant but was gradually decreasing. This change is due to subtle gravitational tugs on the earth from bodies in the solar system. The change is cyclical over a very long period of time, but the trend now is in the decreasing direction. In the late nineteenth century, the astronomer Simon Newcomb worked out a mathematical expression that described the obliquity of the ecliptic as a function of time. In recent years, astronomers have slightly improved on Newcomb's formula.

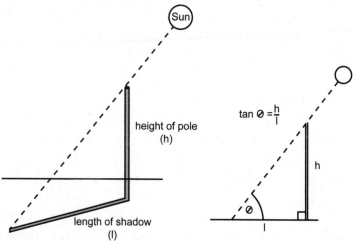

Figure 2.4. The use of a gnomon.

The azimuth (the angle measured from north along the horizon) of sunrise or sunset on the solstices depends upon one's latitude and the obliquity of the ecliptic. There are megalithic structures, such as Stonehenge, that appear to have alignment with the rising or setting sun on one or both solstices. If these alignments truly are with the solstices, then the constructors of the megaliths could have had some understanding of the obliquity of the ecliptic. However, it is more likely that they laid out the orientation of their structures merely by sighting the directions of sunrise or sunset on the solstices. At any rate, it is possible, at least theoretically, to backward engineer and estimate the obliquity of the ecliptic from the measured azimuths of the megaliths. Dodwell collected measurements of some of these structures. He also gleaned measurements of the obliquity of the ecliptic from various ancient sources. These ancient structures do not precisely align with the obliquity of the ecliptic today. The very slight mismatch between the alignment and the obliquity of the ecliptic is attributed to the change in the obliquity of the ecliptic since.

Dodwell thought that his data showed a dramatic departure from Newcomb's formula at earlier epochs. From this, Dodwell concluded that there was a dramatic event that altered the earth's tilt, an event that he dated to the Flood. However, the most important datum for Dodwell's hypothesis is his earliest, the Karnak Temple Complex in Egypt. In the late nineteenth century, one researcher noticed that the alignment of one hall was close to the setting sun on the summer solstice. A subsequent survey showed that the alignment was not exact and hence was due to chance. Keep in mind that without any written record of the purpose of the alignment of any megalith, it is conjecture that the alignment is with the rising or setting sun on a particular date. Today it is generally thought that alignment at Karnak was dictated by the direction of the Nile River's flow. If we remove the inferred measurement of the obliquity of the ecliptic at Karnak, Dodwell's hypothesis is greatly weakened. Therefore, there is no compelling reason, either from biblical or physical evidence, to believe that in the past the obliquity of the ecliptic was dramatically different from what it is today.[4]

---

[4] For more information, see https://answersingenesis.org/astronomy/earth/an-analysis-of-the-dodwell-hypothesis/.

How Do We Tell Time?

The reckoning of time always has been a major purpose of astronomy, and it continues to be today. There are three natural units of time: the day, the month, and the year. These are natural units because they are based on cycles established by astronomical bodies—the rotation of the earth, the orbit of the moon, and the revolution of the earth. The other units of time, the second, minute, hour, and even the week, are not based upon natural rhythms of the world. God ordained that the heavenly bodies be "for days and years" in Genesis 1:14–19. This establishes a purpose for heavenly bodies as useful for time measurement; therefore, the natural units of time amount to a God-given standard. However, God has ordained one non-natural unit of time, the week. The first explicit mention of this was the giving of the fourth commandment in Exodus 20:11. It reads,

> For in six days the LORD made heaven and earth, the sea, and all that is in them, and rested on the seventh day. Therefore the LORD blessed the Sabbath Day and made it holy.

However, since this refers back to the Genesis 1 creation account, one could argue that the ordination of the week goes back to the very beginning of the Bible and creation. How did the Hebrews at Sinai know which day of the week was the Sabbath? There is no indication that God had to instruct them which day the Sabbath was. Nor is there any hint in the giving of the Law that the week was a novel concept. Rather, it appears from the narrative that God was imparting new significance to a week and a particular day of the week (the Sabbath) that the ancient Hebrews were already familiar with. Today people conform to the pattern of a seven-day week without ever questioning the origin of the week. It is thus possible—even likely—that from Adam forward people observed the seven-day week much as we do today.

Unlike the week, minutes, seconds, and hours do not appear to be God-ordained in any sense. Rather, they likely came about by human convention. However, just because this subdivision of time is not God-ordained, it does not follow that this convention is wrong. We are not sure where the practice of seconds, minutes, and hours came from, but they most likely are of Babylonian origin. Minutes and seconds are base 60, and the Babylonians appear to have introduced base six and base 60 reckoning. For instance, the Babylonians introduced the concept of dividing a circle into 360° and subdividing degrees into 60 minutes, with each minute containing 60 seconds. The division of the day and night into 12 hours each or a total of 24 hours is both base two and base six. The eventual division of the hour into minutes and seconds apparently came much later and was instituted in parallel to the manner that degrees were subdivided.

The modern convention of reckoning time of day is to express it in terms of the sun's position with respect to the *celestial meridian*. The celestial meridian (sometimes simply referred to as the *meridian*) is an imaginary line running north-south in the sky and passing through the *zenith*, the highest point in the sky. As the earth rotates, roughly once each day astronomical bodies, including the sun, will cross the celestial meridian, an event we call a *transit*. When the sun transits, we say that it is *noon*. We define the *hour angle* to be the angle that an object makes with the celestial meridian, measured westward. We could express the hour angle in degrees, but for timekeeping purposes it is more useful to express the hour angle in hours, minutes, and seconds of time. A complete revolution is 360°, or expressed in time, 24 hours. Dividing 36 by 24, we find that each hour is 15°. We subdivide each hour into 60 minutes and each minute into 60 seconds. The sun's hour angle is how long ago the sun transited, or how long it has been since noon. Therefore, we can define the time as the hour angle of the sun. Note that hour angle ranges from zero hours to 24 hours, so noon corresponds to zero hours, zero minutes (00:00), while midnight corresponds to 12 hours, zero minutes (12:00). Strictly following this convention, the day would begin at noon, not midnight, which is our normal practice. To remedy this, we add or subtract 12 hours to the hour angle of the sun to get the time, expressed in terms of 24 hours, such as with military time. If the sun's hour angle is less than 12:00, we add 12 hours; if the sun's hour angle is greater than 12:00, we subtract 12 hours.

With this definition, it is relatively easy to construct a sundial. A sundial consists of a gnomon that casts a shadow onto a scale marked off with hours and minutes. The location of the shadow indicates the time on the scale. There are many different designs for sundials, and some sundials can be beautiful works of art. If you have checked the time indicated on a sundial, you probably have noticed that it does not indicate the same time as your watch. There are three reasons for the discrepancy. First, the sun does not keep perfect time.

The time from one solar transit to the next is not always exactly 24 hours, but rather the sun usually runs a few seconds fast or slow. This is because the earth moves approximately $1/365$ of the way around the sun in 24 hours, making the sun take nearly four minutes longer than the stars to go from one transit to the next.[5] Also, due to the earth's elliptical orbit around the sun, this motion each day is not uniform. More importantly, due to the obliquity of the ecliptic, while the sun's apparent motion is along the ecliptic, the rotation of the earth is motion parallel to the equator. Therefore, near the equinoxes, the effect of the earth's orbital motion is foreshortened, but near the solstices, the earth's orbital motion is stretched. Astronomers define the *apparent solar time* as the hour angle of the sun (± 12 hours). Sundials generally keep apparent solar time. Astronomers define *mean solar time* as the hour angle of the *mean sun*, the mean sun being a fictitious sun with these irregularities smoothed out. The mean sun transits every 24 hours, but the real, apparent sun, can run a little more than 15 minutes fast or slow. The difference of the two is called the *equation of time*.

The second reason the time on a sundial generally does not agree with the time on your watch is that a sundial keeps *local time*, while our clocks keep *standard time*. Because the earth is spherical and spins eastward, local time is a function of one's longitude. The local time is four minutes earlier for each degree of longitude westward one travels. For much of history, transportation was slow, normally taking a couple of days to travel just one degree of longitude, so this time difference was not appreciable. However, with faster modern transportation, this became a problem. In the mid-nineteenth century, the railroad industry saw the need to establish time zones, and by the end of the century, many locations had adopted them. Ideally time zones are 15° (one hour) wide and centered on a standard meridian. The standard meridians are measured from the prime meridian in Greenwich, England. The Eastern Standard time zone of the Unites States is based upon the 75th meridian, Central Standard on the 90th meridian, Mountain Standard on the 105th meridian, and Pacific Standard on the 120th meridian. For each degree of longitude that a location is off from the standard meridian, there is a difference of four minutes between local and standard time. For instance, on

---

[5] This leads to the definition of the sidereal day and sidereal time, both being about four minutes shorter than the solar day. Astronomers find sidereal time useful, but it has little application elsewhere.

the 80th meridian, local noon and midnight will be at 12:20, while on the 70th meridian, local noon and midnight will be at 11:40.

The third reason for disagreement between the time on your watch and the time read on a sundial is the application of daylight saving time. In early March in most of the United States[6] we arbitrarily add an hour of time to our clocks and subtract an hour in early November. During these "summer months" we supposedly save something by doing this, but it is not clear what that is. Ideally, time zones are 15° wide, but practical considerations alter this. For instance, dividing major population centers would prove difficult, so time zone boundaries have been moved to accommodate this. However, many people prefer extra hours of daylight later in the day, so the boundary of the Eastern Standard time zone has moved progressively westward over the years. For instance, where I live in Northern Kentucky, local noon and midnight are at 12:39. During most of year when daylight saving time is in force, local noon and midnight are at 1:39. For much of June and early July, sunset is after 9:00 PM. This means that astronomy programs cannot start much before 11:00 PM.[7]

The relatively late introduction of the hour as a unit of time is indicated in Scripture, for while the hour as a unit of time occurs often enough in the New Testament, it does not appear at all in the Old Testament. The word *hour* appears five times in the King James Version of the Old Testament, but those occurrences are late, and they do not refer to a definite unit of time, but rather convey the sense of immediacy. All five occurrences are in the book of Daniel. This is significant, because the setting of Daniel's account is Babylon, where the concept of the hour probably originated. Therefore, it is fitting that an Aramaic word (שָׁעָה)[8] is used. Apparently, the ancient Hebrews did not have a word for hour. If the ancient Hebrews had a way of subdividing the day into smaller units of time, it is unknown to us. The ancient Hebrews probably adopted the concept of the hour during the Babylonian captivity.

---

[6] The exceptions are Arizona (except for the Navajo nation) and Hawaii.

[7] As an astronomer, I have found daylight saving time to be bothersome, so I have taken to calling it "daylight stupid time." In teaching astronomy classes at the university, I usually would let my students know what I thought of this, and sometimes, to my delight, they would argue with me. In the ensuing discussion, I would suggest that we leave our clocks alone and simply do everything an hour earlier. That is, instead of starting class at 8:00 AM, start at 7:00 AM. One of my students once protested that that was too early!

[8] Hebrew and Aramaic are closely related Semitic languages, and they share a common alphabet.

The first two appearances of the word *hour* in the Old Testament (Daniel 3:6, 15) are in the account of Nebuchadnezzar's golden image. These two verses record the command that anyone who did not worship the image was to be thrown into a burning fiery furnace "that same hour." However, newer translations, such as the English Standard Version, the New American Standard Bible, New International Version, and even the New King James Version, render this "immediately" rather than "that same hour." The next two appearances of the word *hour* in the King James Version (Daniel 4:19, 33) are in the context of Nebuchadnezzar's dream about the tree and the fulfillment of the dream. Daniel 4:19 in the King James Version records that Daniel was astonished "for one hour." However, the New King James Version and the New International Version state that Daniel was astonished or perplexed "for a time," while the New American Standard Bible and the English Standard Version record that Daniel was appalled or dismayed "for a while." The final appearance of the word *hour* in the Old Testament (Daniel 5:5) is in the account of the supernatural hand writing on the wall. While both the King James Version and the New King James Version say that the hand appeared "the same hour," the English Standard Version says that the hand appeared "immediately," and the New American Standard Bible and the New International Version say that the hand appeared "suddenly." The fact that newer translations consistently translate שָׁעָה with a word other than "hour" suggests that the intended meaning is immediacy rather than a definite unit of time.

In contrast, the word *hour* occurs 85 times in the King James Version of the New Testament. In the majority of cases, the surrounding context clearly shows that the Greek word is used in the same sense as the Aramaic word in the book of Daniel, in that it does not refer to a definite unit of time. For instance, the term "same hour," meaning right away, appears 11 times in the King James Version of the New Testament. "That hour" appears five times, "very hour" appears twice, and "selfsame hour," "present hour," and "every hour" appear once each. In another non-literal sense, in the Apostle John's gospel the phrase "his hour," referring to Jesus' time, appears three times. Similarly, John 16:21 refers to a woman's time of delivering a baby as "her hour." In His declaration that no one will know the time of His return, Jesus several times said that no one knows the hour (Matthew 24:36–51; 25:13; Luke 12:35–48). In the Garden of Gethsemane, Jesus asked if His disciples could not stay awake with Him for even one hour (Matthew 26:40; Mark 14:41: Luke 22:59). Given that this was stated as a hypothetical, it is not clear that the one hour here should be treated as a literal hour. Nor is it clear in the four times that the phrase "one hour" appears in the Book of Revelation (Revelation 17:12; 18:10, 17; 18:19) that it should be taken literally. A few hundred years ago, most English speakers understood that the word *hour* did not necessarily refer to $1/24$ of a day. It is a shame that this non-literal use of the English word *hour* has fallen out of use today.

However, in other times that numbers are associated with hours (e.g., John 11:9; Acts 10:3; 19:34; Revelation 8:1), it appears that the meaning is literal. In John 11:9, Jesus rhetorically asked if there were not 12 hours in a day. This conformed to the first century Jewish practice of counting 12 hours during the day and 12 hours during the night. These hours were reckoned sunrise to sunset and sunset to sunrise. This is consistent with the Jewish convention of beginning the day at sunset. However, this is different from the Roman practice of reckoning hours from midnight and noon. We follow the Roman custom today. We reckon the 12 hours of AM from midnight and the 12 hours of PM from noon. Noon marks the middle of the day. The term AM is an abbreviation of the Latin phrase *ante meridiem*, which means *before midday*. Similarly, the term PM is the abbreviation of *post meridiem*, meaning *after midday*. Furthermore, we begin the day at midnight.

In reading the New Testament, we must be careful that we understand which time convention—whether Jewish or Roman—is being used. For instance, in the parable of the laborers in the vineyard (Matthew 20:1–15), the owner of the vineyard hired workers at the third, sixth, ninth, and eleventh hours.[9] Since this people were working outside, this obviously refers to the Jewish time convention. Today, we would say that this was at 9 AM, noon, 3 PM, and 5 PM. In similar manner, at Pentecost the Apostle Peter protested that the miracle that people experienced could not be attributed to drunkenness, because it was only the third hour of the day (9 AM). Acts 23:23 refers to the third hour of the night, which would correspond to 9 PM. It is believed that the other three time references in the Book of Acts (Acts 3:1; 10:9, 30) also conform to the Jewish reckoning of time.

---

[9] This passage has given rise to the English idiom, the eleventh hour, referring to late in the day, or just before a deadline.

The only time references in two of the synoptic gospels are related to the Crucifixion. According to Mark 15:25, Jesus was crucified at the third hour (9 AM). Mark 15:33–34 and Luke 23:44 state that there was darkness from the sixth hour to the ninth hour (noon to 3 PM). The other synoptic gospel concurs with the timing of the darkness (Matthew 27:45). Furthermore, Mark 15:34 records that Jesus died at the ninth hour. However, John 19:14 records that Pilate delivered Jesus over to be crucified at the sixth hour. Ordinarily, this would be noon, which was three hours after Jesus was on the cross, according to the synoptic gospels. Is this a contradiction? Unlike the other New Testament references to specific times, John apparently was using the Roman convention of reckoning time from midnight rather than sunrise, so this was 6 AM. Why did John do this? It is generally conceded that John wrote his gospel later than the other three gospels, after the fall of the Jerusalem and the Temple's destruction in AD 70. John probably wrote his gospel in Ephesus, far from Judaea. Given the late date and location of writing, it is likely that John was writing largely to a Gentile audience. With neither a Temple nor a large presence of Jews in Judaea, it made more sense, when writing to a Gentile audience, to use the non-Jewish system of reckoning time. While Mark's gospel may have been written to a predominantly Gentile audience, he apparently used the Hebrew convention of time reckoning for the time of the Crucifixion.

There are only two other time references in the gospel of John. The first reference (John 1:39) records that the first two disciples stayed with Jesus the same day that they met him, for it was the tenth hour. The wording implies that the day was yet quite young, but if this narrative followed the Jewish custom of reckoning time, the time would have been 4 PM, which would have been late in the day. On the other hand, if the Roman custom was observed here, it was but 10 AM, with much of the day still ahead. The remaining time reference (John 4:6) refers to when Jesus spoke to the Samaritan woman at Jacob's well. It was the sixth hour. Many people assume that this followed the Jewish custom, making this noon, the warmest time of the day. However, if John followed the Roman custom here as in the other two instances, the time would have been 6 PM. While not as hot as noon, it would have been at the end of a very long day of walking, and Jesus would have been thirsty. The narrative works with either time convention, but it would be more consistent for John to have used the Roman time convention all three times rather than two times out of three.

References to months and years occur more often in the Bible than references to hours. However, the calendar used in biblical times is very different from the calendar that we use today. The English word for month comes from an Anglo-Saxon word for the moon. This is because the moon's phases were the basis for the month. The moon orbits the earth with respect to the sun once each *synodic month*, about 29½ days. The phases of the moon repeat on this cycle (Figure 2.5). We normally begin lunar phases at new moon, with the moon between the earth and sun. Because the new moon appears so close to the sun in the sky, new moon is not visible. In fact, the moon is not visible at all for two or three days around new moon. One to two days after new moon, the moon travels far enough along its orbit around the earth that it is visible as a thin crescent low in the southwestern sky shortly after sunset.[10] The moon appears as a crescent because the lit half of the moon is facing the sun, and we mostly see the unlit, dark half of the moon, with just a thin sliver of the lit half visible. Each night the illuminated crescent grows, so we call this the waxing crescent phase.

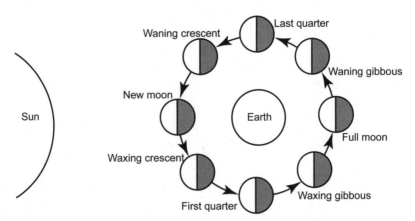

Figure 2.5. Phases of the moon.

About a week after new moon, the moon appears half lit. However, we call this a first quarter moon rather than a "half-moon." The word *quarter* here describes how far the moon has traveled in its orbit, not how the moon looks. Each night after first quarter, the amount of the lit portion of the moon that

---

[10] This is true in the Northern Hemisphere; in the Southern Hemisphere, the waxing crescent moon is in the northwestern sky.

we see continues to grow. This is called the gibbous phase. More specifically, it is called the waxing gibbous. About a week after first quarter, the gibbous phase has grown so that the moon is fully illuminated, making the moon full. After full moon, the phases reverse, with waning gibbous, third quarter, waning crescent, and then finally new moon, which completes the cycle and starts the cycle anew. The waning phases resemble mirror images of the waxing phases. The waning phases are lit on the right, while the waxing phases are lit on the left (this is reversed in the Southern Hemisphere). The waxing phases are visible in the early evening, right after sunset. The waning phases are visible in the early morning, just before sunrise.

The moon is very bright, and hence noticeable, in the night sky, so it makes a very effective time marker. Furthermore, the month is an effective intermediate time measurement between the other two natural time measurements, the day and the year. The problem is that, being 29½ days long, the synodic month is not easily divisible by the day. Most ancient calendars relied upon direct observation for determining the duration of the month, so most months varied between 29 and 30 days. We call such an arrangement a *lunar calendar*. The most common arrangement was to look for the first thin crescent moon visible after sunset following the new moon when the moon was not visible. Today we define new moon astronomically so that new moon is when the sun and moon are as close to the same direction in space using the earth as the reference point. However, in most ancient cultures, new moon would have been defined as the first visibility of the waxing crescent, essentially starting the new phase cycle (and hence the name "new moon"). If inclement weather prevented direct observation of the new moon, then one might think that some months would have more than 30 days. However, with just a few years of observation and modest record keeping, one can learn to anticipate when the new moon ought to have appeared, even in cloudy weather. One algorithm used when cloudy weather intervened simply was to declare a new month after 30 days had elapsed.

Another difficulty arises from the fact that 12 months defined this way amounts to 354 or 355 days, ten or 11 days short of a year. The Islamic calendar is the only lunar calendar widely in use today. On the Islamic calendar, the months, including the holy month of Ramadan,[11] occur about 10 days earlier

---

[11] If there is a third book in this series, an important topic that I intend to explore is who put the ram in the Ramadan-a-ding-dong.

each year. Therefore, the month of Ramadan drifts successively earlier with respect to the seasons over a cycle of about 32 years. One of the primary purposes of a calendar is to keep our time reckoning synchronized with the seasons. This is particularly important for agricultural purposes. Obviously, a strictly lunar calendar is useless for this. Notice that the error is about $1/3$ of a month per year, so the most obvious fix is to add an additional month every three years. We call this additional month an *intercalary month*. Since this kind of calendar attempts to maintain the true lunar month while at the same time synchronizing with the sun, we call this kind of calendar a *lunisolar calendar*.

The Jewish calendar upon which Jewish festivals are based is a lunisolar calendar. With the giving of the Law at Mount Sinai, God instituted the ceremonial year that began in early spring, close to the vernal equinox. Previously, the ancient Hebrews had observed New Year six months later, near the time of the autumnal equinox. The Jews continue the practice of the civil year beginning six months after the beginning of the ceremonial year. The Hebrew term for the Jewish New Year, *Rosh Hashanah*, means *head of the year*. Incidentally, Jewish tradition holds that the first day of the Creation Week was *Rosh Hashanah*. The first day of each month on the Jewish calendar coincides with the first visibility of the crescent moon. Since in Hebrew reckoning the day begins at sunset, the first day of the month coincides with the observationally based new moon. Defined this way, the full moon always occurs 14 days later, on the fifteenth day of the month. Two Jewish festivals fall on the fifteenth day of their respective months. Passover was a week-long observance that began on the fifteenth day of the first month on the ceremonial calendar, while the Feast of Tabernacles was another week-long observance that began on the 15th day of the seventh month on the ceremonial calendar (two weeks after *Rosh Hashanah*). Being a lunisolar calendar, the first days of these two festivals also fall on the full moon.

When are intercalary months added? There is evidence that the ancient Hebrews based the need for an intercalary month on observation of the barley harvest. Grain crops typically were sown in autumn and grew throughout the rainy winter into spring. Barley has a shorter growing season than wheat and other grains, so it would ripen first, usually in late winter and early spring. The Feast of Firstfruits was on the second day of the Passover feast. This required a wave offering of the first sheaves from the barley harvest. Examination of the barley crop toward the end of the twelfth month of the ceremonial calendar (late winter—February/March) would indicate whether the barley harvest would be ready a little more than two weeks later for the firstfruits offering during Passover. If the barley crop required additional time, then an intercalary month was inserted. However, if it appeared that the barley would begin to mature at the appropriate time, then no intercalary month was inserted. The name of the first month on the Hebrew calendar, *Aviv*, has the meaning of the stage of development of a grain crop that was beginning to mature. The second major festival was the festival of weeks or harvest. It was 50 days after Passover.[12] This festival coincided with the conclusion of the grain harvest. The third major festival, the Feast of Ingathering or Tabernacles, was six months after Passover. This festival coincided with the harvest of non-grain crops. Since all three of these festivals were tied to the harvest of agricultural products, it was vital that they remain synchronized with the seasons.

Sometime after the destruction of the Temple in AD 70, the Jews abandoned the observational method of deciding on when to add an intercalary month. This probably was related to the Diaspora with different climates where the Jews lived and a lack of centralized authority that revolved around the Temple. Hillel II in the third century generally is credited with the Jews' adoption of the Metonic cycle. The Metonic cycle is a method of intercalation attributed to Meton, a fifth-century BC Athenian who discovered it, though Babylonian astronomers already knew about it. Proper intercalation requires the insertion of seven months over a 19-year period. The Metonic algorithm the Jews adopted requires an intercalary month at the end of the third, sixth, eighth, eleventh, fourteenth, seventeenth, and nineteenth years of a 19-year cycle. This method of intercalation does not require any further adjustment for over three millennia.

This sort of calendar is very different from the one that we use today—the Gregorian calendar. The roots of our calendar go back to ancient Rome. The early Romans had a lunar calendar, but eventually the Romans adopted a lunisolar calendar. The lunisolar calendar required intercalary months, but the process of deciding when to insert a month was plagued with corruption, so intercalary months were not added methodically, which led to much chaos.

---

[12] We get the equivalent Greek word for this festival, Pentecost, from the word for 50.

Eventually, Julius Caesar instituted a calendar reform that took effect in 44 BC. The Julian calendar was a *solar calendar* that was tied to the length of the *tropical year* rather than the synodic month. The tropical year is the orbital period of the earth with respect to the vernal equinox. Using the tropical year as a basis for the calendar guarantees that the calendar will remain fixed with respect to the seasons without the need for an intercalary month. The Julian calendar required dispersing 11 extra days over the 12 months to reach 365 days in a year, resulting in months, generally, of 30 and 31 days we now have. Thus, this change amounted to an abandonment of a lunar-based month. The tropical year is nearly ¼ day longer than 365 days, so the Julian calendar also introduced a leap day every fourth year. On the original Roman calendar, February was the twelfth month of the year, so that is why February is the month when we add the leap day.[13] The Egyptians had adopted a solar calendar about two centuries earlier, which may have influenced the decision for the Julian calendar reform.

Contrast this sort of calendar to a lunisolar one. Passover begins on the 15th of day of the month of *Aviv*. This always is a full moon. The month of *Aviv* begins in March or April, so Passover can fall on almost any date in April and occasionally in late March on our calendar. It appears that Passover bounces around on our solar calendar, but it remains the same date on a lunisolar calendar. For that matter, the date of Resurrection Sunday bounces about on our calendar, so how is the date of Resurrection Sunday determined? We know that Jesus rose on a Sunday morning. We know that it was immediately following Passover. Hence, if we wish to celebrate the resurrection of Jesus Christ once per year, it ought to be on the Sunday immediately following Passover. Passover generally is the first full moon after the vernal equinox. At the Council of Nicaea in AD 325, the church adopted this formula: Resurrection Sunday is the first Sunday after the first full moon after the vernal equinox. While this definition often causes Resurrection Sunday and Passover to be within a few days of one another, they sometimes are separated by about a month.

---

[13] September, October, November, and December are the ninth, 10th, 11th, and 12th months of our calendar, but they were the seventh, eighth, ninth, and tenth months on the Roman calendar. The names of these months are from Latin. Anyone who has studied a Romance language can see that the names of these months contain the numbers seven, eight, nine, and ten.

However, there was another wrinkle. By methodically adding a leap day every four years, the Julian calendar effectively defined the year to be precisely 365.25 days. But the tropical year (the cycle over which the seasons repeat) is slightly less than 365.25 days. This introduced an error of three days every four centuries. This caused the months to slip earlier in the year, so that by the sixteenth century, there was an error of ten days. The vernal equinox was occurring around March 11, rather than March 21 as it did at the time of the Council of Nicaea. Furthermore, Resurrection Sunday was happening earlier in the year too. To correct this, Pope Gregory XIII established a revision to the Julian calendar in 1582. This new Gregorian calendar revised the method of adding leap days. Even century years, except those divisible by 400, were now common years, years without a leap day. Therefore, the years 1700, 1800, and 1900 were not leap years, though on the Julian calendar they would have been. However, the years 1600 and 2000 were leap years. The Gregorian calendar will keep the calendar synchronized to the seasons to within a day for thousands of years.

There was a desire to restore the calendar as it existed at the Council of Nicaea in AD 325. This required eliminating ten days from the calendar. Therefore, in 1582, October 4 was immediately followed by October 15. This was very controversial, and many people demanded their ten days back. If that weren't confusing enough, the Gregorian calendar reform attempted to standardize the observance of New Year. Different countries, and even different regions within countries, had different customs regarding the start of the year. One of the more common dates for New Year's Day was March 25, the date of the vernal equinox when the Julian calendar took effect. March 1 was another date for New Year, but April 1 was popular elsewhere. Still other countries observed the Roman custom of New Year's Day being January 1. The Gregorian calendar reform chose the Roman custom of January 1 for the observance of New Year's Day.

Adding to the confusion was the fact that the Gregorian calendar reform came about just a few decades after the Protestant Reformation began. Roman Catholic countries readily adopted the new calendar, but Protestant and Orthodox countries did not. For instance, Britain and its colonies did not adopt the Gregorian calendar until 1752. Since the error of the Julian calendar by that time had accumulated an additional day, this change required eliminating 11 days from the calendar. The last few holdouts adopted the Gregorian calendar in the early 20th century. These were Orthodox countries, such as Russia and Greece. This required dropping 13 days from the calendar. Russia adopted the Gregorian Calendar in 1918, the year following the Communist revolution. The Communists called their revolution "the October Revolution," because of the month in which it occurred. However, it was October only in Russia and a few other places—everywhere else it already was November.

Was the Year Once 360 Days Long?

Related to our discussion of calendars is a common belief among biblical creationists that the year once was 360 days long, and that the year consisted of 12 30-day months. I critiqued this idea in some detail in Chapter 10 of *The Created Cosmos: What the Bible Reveals about Astronomy*. Therefore, I will only briefly discuss it here, emphasizing a point that I did not make there.

A key part of the argument for an original 360-day year comes from what many people see as a prophetic year of 360 days in Scripture. Mind you, the Bible does not explicitly teach this, but rather it is an inference that many people take from certain passages in the book of Daniel and the book of Revelation. Daniel 7:25, Daniel 12:7, and Revelation 12:14 contain the unusual phrasing, "time, times, and a half a time." The first time is thought to be a year, the times is thought to be two years, and the half a time to be a half year, for a total of 3½ years. Revelation 11:2 and Revelation 13:5 mention a period of 42 months. Given that a year has 12 months, it is inferred that the 3½ years and 42 months are the same duration. Furthermore, Revelation 11:3 (immediately following Revelation 11:2 and its mention of 42 months) and Revelation 12:6 indicate time periods of 1260 days. Again, this is at least approximately equal to 42 months or 3½ years, implying to some people an exact equivalence, and hence a prophetic year of 360 days. Daniel 9:20–27 is Daniel's prophecy of 70 weeks, with each week generally understood to correspond to seven years, for a total of 490 years. While the text of Daniel 9 does not identify the duration of the time periods involved, proponents of the 360-day prophetic year argue that the aforementioned verses in Revelation and elsewhere in Daniel imply that the prophetic year is 360 days. There is also the historical argument, as the date of the Crucifixion is almost exactly 483 schematic (360 day) "years" after the decree to rebuild Jerusalem, as prophesied by Daniel. However, because Daniel does not use the term "year" but simply employs the term "seven/week," we are not constrained to call the "prophetic year" a year. *It is simply a set time period.* Notably, this scheme works only if one acknowledges that the prophetic and actual years have slightly different lengths.

Sir Robert Anderson (1841–1918) popularized a common theory about the prophetic year in his book, *The Coming Prince.* Anderson reasoned that according to Daniel 9:24 there were 69 weeks between Artaxerxes' authorization to rebuild the walls of Jerusalem (cf. Nehemiah 2:1–9) until the Crucifixion (i.e., the "cutting off") of Jesus the anointed Messiah. According to this thesis, we must compute the duration of the 69 weeks in days: $69 \times 7 \times 360 = 173{,}880$ days. However, to find the duration in terms of the Julian calendar,[14] we must divide this duration in days by the length of the year on the Julian calendar: $\frac{173{,}880}{365.25} = 476$ years, 21 days. Nehemiah 2:1 records that it was in the month of *Nisan (Aviv)* during the twentieth year of the reign of Artaxerxes that he gave the decree to rebuild the walls of Jerusalem. The reign of Artaxerxes generally is understood as having begun in 465 BC, which would place the decree to rebuild the walls of Jerusalem in 445 BC. Adding 476 years to this (and knowing that there was no year zero), we arrive at a date of AD 32. This is very close to the date of the Crucifixion of Jesus.

At this point, Anderson's theory assumes several different forms. Holding to Anderson's original work, some insist that the Crucifixion occurred in AD 32. Making certain assumptions about when during the month of *Nisan* that Artaxerxes made the decree and other assumptions about the calendar, one may conclude that either Jesus rode into Jerusalem exactly 173,880 days (483 360-day years) after the decree, or that the Crucifixion was exactly 173,880 days after the decree. Either way, according to the theory this would be a literal fulfillment of Daniel's prophecy. However, most scholars consider the years AD 30 or AD 33 to be more likely dates of the Crucifixion and hence the triumphal entry into Jerusalem. Xerxes died in late December of 465 BC, and Artaxerxes assumed the throne immediately. However, because the Persians held to the accession year system in dating the reigns of their rulers, Artaxerxes' first year on the throne was reckoned from *Nisan* 464 BC to *Nisan* 463 BC. This means

---

[14] Use of the Julian rather than the Gregorian calendar is necessary here because it is customary to express dates prior to the adoption of the Gregorian calendar in 1582 as Julian dates. However, it is of little consequence because the difference is a matter of only a few days.

that his twentieth year extended into the Spring of 444 BC, which is regarded by the best biblical chronologists as the date for the issuing of the decree to rebuild Jerusalem. An addition of 476 years from that point lands us in AD 33.

Some have attempted to use this sort of reasoning as a demonstration of the inspiration of Scripture: only if the Bible is divinely inspired could such a prophecy made five centuries earlier be fulfilled to the day. Obviously, from the biblical and secular historical data there is not enough information to reach such a precise conclusion. However, the approximate concordance of the dates is compelling. The salient point to make here is that anyone who uses this argument tacitly agrees that the 360-day prophetic "year" is a motif rather than a reflection of some hypothetical real calendar of the past. Hence, it is inconsistent to use the argument to maintain that the year once was exactly 360 days long.

How Long Did the Flood Last?

Related to the question about a 360-day year is the question of the duration of the Genesis Flood. Genesis 7:11 records that the Flood began on the seventeenth day of the second month of the six hundredth year of Noah's life. Genesis 8:14 records that the earth had dried out on the twenty-seventh day of the second month of the six hundred and first year of Noah's life. If one were naively to assume that the calendar in use was the same as today's Gregorian calendar, then the Flood's duration was 375 days. However, as we have already seen, the Gregorian calendar is a relatively modern convention, and it is extremely unlikely that such a calendar was in use at the time of Noah. What kind of calendar did Noah use? Probably a more important question is what calendar did the ancient Hebrews use, because Moses is credited with authorship of the Pentateuch, which includes the Flood narrative of Genesis 6–9. It is important to note that the Hebrews understood what the Flood narrative meant at the time the narrative was written.

The biblical chronologist Archbishop James Ussher was of the opinion that the ancient Hebrews observed a solar calendar. In support of his position, Ussher argued in the preface of *Annals of the World* (1650) that "It cannot be proven that the Jews used a lunar calendar until after the Babylonian captivity." This is a poor argument. This amounts to saying that since A cannot be proven, not A must be true. However, one easily can reverse this—since not A cannot be proven, then A must be true. Obviously, this reasoning is spurious. Even if the ancient Hebrews did use a solar calendar, we have no knowledge of its structure. Hence, any attempt to establish dates on this hypothetical calendar, as Ussher did, is conjecture.

The assumption of a 30-day month, 360-day year calendar results in a Flood duration of 370 or 371 days. The 370-day figure results from merely taking the difference between the beginning and ending of the Flood given in Genesis 7:11 and Genesis 8:14. However, the ancient Hebrews normally counted the first and last day of a duration, so this would result in 371 days. The widespread belief in a 371-day Flood among recent creationists probably is due to the influence of John C. Whitcomb and Henry M. Morris. In their foundational book *The Genesis Flood*, published in 1961, Whitcomb and Morris endorsed this position. They did not present a calculation for this, but rather Whitcomb and Morris reproduced a table that they referenced from a commentary on Genesis by E. F. Kevan in *The New Bible Commentary*, published in 1953. Interestingly, Kevan did not endorse this figure as the length of the Flood. Instead, Kevan preceded his table with the statement that the Flood chronology may be constructed this way, but he followed the table with the remark that this result assumes a 30-day month. He further noted that the synodic month actually is 29½ days, and from this Kevan concluded that the Flood's duration was 365 days. Hence, Kevan actually endorsed the view that the Flood lasted exactly one year. The manner in which Kevan presented his table suggests that the idea of the 371-day Flood duration preceded him in the literature, and indeed it did. It is clear that belief in the duration of 371 days for the Flood relies upon the assumption that the Flood account of Genesis 6–9 was based upon a calendar that used 12 30-day months. As previously mentioned, Ussher believed that the ancient Hebrew calendar was of this type, so he thought that the Flood's duration was 371 days. Ussher's belief probably influenced Whitcomb and Morris.

Again, with Mosaic authorship of the Pentateuch and of the Flood narrative, the crucial question is not what possible calendar Noah may have used, but what calendar Moses likely used. Most Bible scholars believe that the ancient Hebrews always observed a lunisolar calendar, though some believe that the Hebrews adapted their calendar from the Egyptians during their bondage. There are several reasons for this. When did the year begin in the Old Testament? There are indications that the calendar originally followed in the

Old Testament placed New Year near the time of the autumnal equinox (Exodus 23:16; 34:22). Even today, the Jewish New Year, *Rosh Hashanah*, is a new moon close to the autumnal equinox. This begins the civil year. However, at the first Passover, God instituted a ceremonial calendar with its first month near the vernal equinox (Exodus 12:2; 23:15), six months prior to Rosh Hashanah. The Passover is explicitly stated to be in the month of *Aviv* (אָבִיב). There are only three other Hebrew names for the months mentioned in the Old Testament: *Ziv* (זִו, the second month; 1 Kings 6:1, 37), *Ethanim* (אֵתָנִים, the seventh month; 1 Kings 8:2), and *Bul* (בּוּל , the eighth month; 1 Kings 6:38). The Hebrews eventually abandoned their Hebrew names of the months for Babylonian names, probably during the Babylonian captivity. These are the names used today in the Hebrew calendar, but the only scriptural reference to any of those names is Esther 3:7, where the Babylonian name *Nisan* (נִיסָן) is used instead of *Aviv*.

How long have the Hebrews followed this calendar? We do not know. Credit for codifying the current Hebrew calendar usually goes to the twelfth century AD scholar Maimonides (1135–1204). How much did he contribute to this calendar? At the very least, Maimonides instituted the dating of the Hebrew calendar from the creation of the world rather than from the destruction of the Temple, as had been done previously. However, it is unlikely that Maimonides dramatically altered the calendar. The Hebrews had been fastidious about maintaining the integrity of the law and their customs. The dates of festivals, such as Passover, are fixed on a lunisolar calendar. If Maimonides had changed the Hebrew calendar substantially, say from solar to lunisolar, there would have been much opposition and discussion, because this would have altered the days upon which the various feasts fell. No, this aspect of Hebrew custom strongly argues that Maimonides at most tweaked the calendar that had been in use for some time. Codifying normally means to clearly write out those practices already in existence. How far into the past can we extrapolate the current Hebrew calendar? In the absence of clear evidence that the Hebrews dramatically changed their calendar at some time in the past, it is compelling to extrapolate its use indefinitely into the past. Any dramatic change in the calendar would have greatly altered the very days on which Passover and other feasts were observed. From what we know of the extreme commitment that the Hebrews have had to preservation of tradition, this is almost unthinkable.

Moses instructed that the priests were to offer a special sacrifice at the beginning of each month (Numbers 28:11–14). In many other passages, such as the description of the dedication of the Temple (2 Chronicles 2:4), it is recorded that these sacrifices were at the new moon. While these passages read differently in English translations, there is no difference in the original language. The Hebrew word used in each instance, *hōdeš* (חֹדֶשׁ), means new moon. If a lunar or lunisolar calendar were used, the meaning would be synonymous with the first day of the month, because the first of the month and the new moon would coincide. However, this usage does not make sense if a solar calendar were used.

Given these considerations, it is very likely that the Hebrews observed a lunar or lunisolar calendar as far back as the time of the Exodus. Of course, this does not guarantee that such a calendar was followed as far back as the Flood, but in the absence of any clear evidence that there was at some point a change in calendars, it is reasonable to infer that this was the kind of calendar used in the account of the Flood in Genesis 6–8. Furthermore, Moses wrote the Pentateuch a millennium or more after the Flood. Even if the lengths of the month and year had changed at the time of the Flood, use of some alleged pre-Flood calendar would have made no sense to the readers of what Moses wrote. Rather, the readers would have understood the writings of Moses in terms of the calendar that they understood. Hence the Flood account cannot be used in support of an alleged pre-Flood calendar that was fundamentally different from the post-Flood calendar.

Some might argue that 30-day periods mentioned in the Old Testament indicate that the ancient Hebrews observed months that were 30 days in length. For instance, when Aaron and Moses died, the Israelites mourned 30 days for them (Numbers 20:29 and Deuteronomy 34:8). Also, pagan kings evidently had some official protocols connected to a 30-day period (e.g., Esther 4:11 and Daniel 6:7, 12). However, there is no mention of a month in such passages, so to assume that this amounts to a definition of the month and then to conclude that all the months of the year were 30 days is reading into the text an interpretation that is not justified. The true length of the month rounds to 30 days, so the use of 30 days in such passages suggests a period of approximately one month, but it implies no more than that. By way of analogy, if 1000 years from now some researcher found ancient records showing that in America in the twenty-first century the U.S. Flag Code called for flags to fly at half-staff during a 30-day mourning period when a current or former President died, that would not be evidence that twenty-first century Americans used a calendar with 12 30-day months. Contracts and other legally binding documents frequently include 30-day intervals or multiples of 30-day intervals. These periods approximate the length of the months on our calendar, so obviously they are meant to roughly translate into months, but this use hardly requires that the months in the Gregorian calendar are all 30 days long. In similar manner, Old Testament passages that mention 30-day periods do not prove that the months on the ancient Hebrew calendar were all 30 days long.

Nor is there biblical or extra-biblical evidence for a 360-day year. Calendars with 30-day months were rare among ancient cultures. One exception is the ancient Egyptian calendar, though it postdated the Flood. But the ancient Egyptian year did not have 360 days, because they added five days at the end of the year to produce a 365-day year, so the Flood's duration would then have been 375 or 376 days, not 371 days.

In days, how long was the Flood? Genesis 7:11 and Genesis 8:14 give the exact dates of the beginning and end of the Flood, revealing an elapsed time of 12 months and 10 or 11 days, depending upon how one might count the first and last days. If, as seems likely, the Flood narrative is expressed in a lunisolar calendar, 12 months normally would be 354 days (but occasionally 355 days). If we add 11 days to 354 days, we get a duration of 365 days, one full year. In his commentary, Kevan clearly endorsed this figure for the Flood's duration, as did the Cassuto in his commentary on Genesis.[15] Cassuto further points out that this nicely explains why the LXX differs from the Masoretic text in Genesis 7:1.[16] Instead of the Flood beginning on the seventeenth day of second month, the LXX gives the date as the twenty-seventh day of the second month. The LXX was translated in Alexandria. Jews living in Egypt at the time of the translation of the LXX likely were accustomed to the 365-day year that the Egyptians employed, and this change in the text reflected their understanding of the Hebrew culture that the Flood's duration was exactly one year. Indeed, the case for the Flood lasting exactly one year is good.

---

[15] Umberto Cassuto, *A Commentary on the Book of Genesis, Part Two: From Noah to Abraham*, translated by Israel Abrahams (Jerusalem: The Magnes Press, 1964), 45.

[16] Ibid., 113–114.

CHAPTER 3

He Formed It to Be Inhabited:
The Earth as a Planet

Geology is the study of the earth, so you may wonder why a book on astronomy would include a chapter on the earth. The earth is a planet, just one of eight planets orbiting the sun. Since planetary science is a sub-discipline of astronomy, the study of the earth is an appropriate part of astronomy. We have an advantage in studying the earth over the other planets because we live here and hence we can probe the earth directly. Furthermore, there are similarities between the earth and other planets. *Comparative planetology* is the study of planets by comparing their similarities and contrasting their differences. By studying the earth, we can draw some conclusions that may be true about other planets, both within the solar system and perhaps orbiting other stars.

What Is the Earth Like Inside?

The earth is a sphere 8000 mi (12,875 km) in diameter. Knowing the earth's diameter, we can compute its volume. We know the mass of the earth by studying the moon's orbit around the earth. The earth's gravity provides the centripetal force on the moon that results in the centripetal acceleration required to keep the moon in orbit. According to Newton's law of gravity, the acceleration depends upon the distance between the earth and moon and the earth's mass. We know the moon's distance, so from Newton's law of gravity we can determine the earth's mass. Dividing the earth's mass by its volume reveals the earth's density, 5.5 grams per cubic centimeter (g/cm^3). Water has a density of $1.0\,g/cm^3$. The density of rocks on the earth's surface typically are near $3.0\,g/cm^3$.

Rocks do not compress well, so if the interior of the earth were made of the same material as its surface, the earth's density would be close to 3.0 g/cm³. The fact that the earth's average density is far higher than that reveals that the earth's interior must contain material that has significantly higher density than the density of surface rocks, and a density even higher than earth's average density. There are many elements that have density greater than 5.5 g/cm³, but many of those substances are relatively rare in the universe. The only elements that have sufficient density to account for the earth's high average density yet are common in the universe are iron and nickel. Therefore, unless the earth's overall composition is very different from the rest of the universe, there must be much iron and nickel within the earth. The fact that earth's surface rocks are less dense than the average earth density suggests that the earth is *differentiated*. That is, the earth's material has been separated according to density. This means the earth is divided into layers of different density. We would expect then that most of the earth's iron and nickel is near earth's center, in a region that we call the core. Lying above the core is a rocky region called the mantle. The mantle accounts for most the earth's volume. Atop the mantle is the relatively thin crust. While rocky like the mantle, the composition of the crust is different from the mantle.

Is there any other evidence that the earth is differentiated? Yes. There are different types of meteorites, fragments of small solar system bodies that have fallen to the earth's surface. The two basic types are stony meteorites and iron meteorites. This difference in composition suggests that the parent body or bodies of meteorites underwent differentiation as well, with the iron-type meteorites being samples of the core or cores of those parent bodies. The composition of iron meteorites matches the inferred composition of the earth's core. Differentiation probably is a common feature of many bodies in the solar system. As for the mantle, geologists believe that some basalt and most kimberlite/lamproite (diamond pipe) eruptions bring up material from the upper mantle. These rocks are different from crustal rocks, but they fit the proposed composition of the mantle.

However, the model of the earth's interior has much more detail than this. Geologists recognize that the earth's surface is the top of a thin layer that we call the crust. The earth's crust varies in depth between three and 40 miles. The crust generally is thicker over the continents and thinnest over the deepest basins of the ocean. Below the crust is the mantle, a region that encompasses most of the earth's volume (nearly 85%). The mantle is made of rock, though generally it is a type of rock that is a bit different from rocks of the crust. Below the mantle is the core, where most of the earth's iron and nickel are located (Figure 3.1). As the earth's interior is hot, it would be expected that the temperature generally

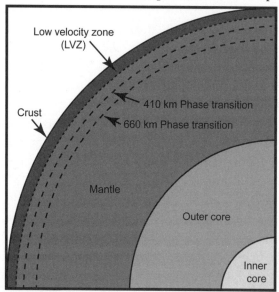

Figure 3.1. Cross section of the earth (including inner and outer core) (Andrew A. Snelling, "Journey to the Center of the Earth." *Answers* 8, no. 3 [2013], 724).

increases with increasing depth inside the earth. This increase in temperature along with an increase of pressure relative to depth have profound effects upon the earth's interior structure.

The earth's core has two distinct layers, the inner and outer cores. The primary difference is that the inner core is solid, while the outer core is molten. This is counterintuitive, as one might expect the inner core to be hotter than the outer core, and so one would think that if any part of the core would be molten, it would be the inner core rather than the outer core. However, the temperature within the core is nearly constant, because the core is made mostly of iron, which is a reasonably good conductor of heat. If the center of the core were appreciably hotter than the outer regions of the core, heat would flow outward relatively quickly to establish something close to the same temperature throughout the core. Still, why would some of the core be molten, while the rest is not? Most people are aware that increased pressure raises the boiling point of a substance. This is how a pressure cooker works—the pressure in the cooker significantly raises the boiling point of water above the normal boiling point so that food cooks more quickly. Less well known is the fact that increased pressure also elevates the melting point of substances. The pressure is greatest at the earth's center. The pressure is so great at the earth's center that the melting point of iron is raised above the temperature of the core, so the iron remains solid there. However, as the pressure in the core decreases with distance from the earth's center, a point is reached in the core where the melting point of iron matches the temperature of the iron, so that the iron is molten from that distance outward.

Seismology provides a wealth of data to support this model. Seismology is the study of earthquakes as the waves generated by them move through the earth. This is similar to how sonograms can image internal structures of the human body without opening the body via surgery. The waves that pass through the earth are P-waves (for primary) and S-waves (for secondary). The P-waves move more quickly, so they are detected before the S-waves are (the words *primary* and *secondary* refer to the order in which the waves arrive at a seismograph). P-waves can travel through both solid and liquid, but S-waves can pass through only a solid. On the other side of the earth from an earthquake's location, there is a large region in which seismographs detect only P-waves, indicating that there is a portion of the earth's interior that is liquid. The region of the S-wave shadow always is symmetrical around the point diametrically opposite to where the earthquake occurs (Figure 3.2). This

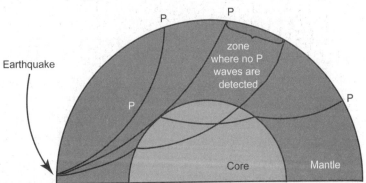

Figure 3.2. Illustration of movement of P-waves and S-waves (and reaching interface) (Andrew A. Snelling, "Journey to the Center of the Earth." *Answers* 8, no. 3 [2013], 72–75).

indicates that the liquid portion of the core is concentric with the earth's center. From the data, seismologists can determine the outer radius of the liquid portion, and this radius is consistent with the radius of the core.

What about the solid inner core? Whenever P-waves reach an interface, they reflect. For instance, faint echoes of P-waves bounce off the boundary between the liquid outer core and the solid mantle. Seismographs can detect these reflections near the location of the earthquake. Once a P-wave penetrates completely through the liquid outer core, it may reflect off the interface between the solid inner core and liquid outer core. Again, seismographs can pick up this echo, indicating that there is a solid inner portion to the core. With these reflections, the entire signal is a bit complicated, but it can be sorted out in a straightforward manner.

Adding to the complexity is the manner in which P-waves and S-waves move through the mantle. Waves that move directly downward (along a radius) from an earthquake move straight. Waves that do not move directly downward initially move along chords of the earth. However, those waves soon end up gently curving upward from their initial trajectories. This is caused by gradual changes in properties of the material with greater depth in the mantle. Part of these changes are due to changes in composition, while others are caused by changes in density induced by pressure. Wave velocity increases with depth, and refraction occurs when waves change speed. Geologists study the way the waves are bent upward to model the interior structure of the earth.

The entire picture that seismology provides is consistent with the model of the earth's interior, including its liquid outer core and solid inner core, both consisting primarily of iron. But there is another line of evidence for the structure of the earth's core: the earth's magnetic field. Magnetic fields are generated by electrical currents in conductors. The presence of ferromagnetic substances, such as iron and nickel, can greatly boost the strength of magnetic fields. The simplest explanation is that there is a movement of material in the liquid outer core. This convection in the outer molten core would include the movement of charge, so that an electrical current exists in the liquid portion of the earth's core. This current generates the earth's magnetic field.

If this model of the earth's magnetic field were correct, then we would expect the field slowly to decrease in strength. This is the nature of most currents—they decay because of electrical resistance. Electrical devices have energy sources, such as a battery, to maintain current in a circuit, but the earth does not have a battery. Indeed, measurements of the earth's magnetic field for the past 180 years show that the field strength is decreasing. This probably is an exponential decay. If we extrapolate this exponential decay of the earth's magnetic field into the past, we find that only a few million years ago the current would have been so great that the heat generated by the electrical resistance would have melted the earth. Obviously, this cannot have been the case. Of course, if the earth is only thousands of years old, this is not a problem. How do those who believe in 4.6 billion-year-old earth respond to this? They posit that the earth's magnetic field goes through reversals. The earth's gross dipole magnetic field allegedly fades to zero, but then it regenerates with an opposite polarity, builds to some maximum intensity, and then fades back to zero, whereupon it regenerates in the reverse direction once again.

Is it possible that the earth's magnetic field has undergone and continues to undergo multiple reversals? Perhaps, but such a thing would require a mechanism to regenerate the current each time after resistance dissipates the current to zero. There is no such known mechanism. Those who believe in an age for the earth on the order of billions of years appeal to a subtle signature of magnetic field orientations in some volcanic rocks as evidence of magnetic field reversals. When a rock cools and hardens in the presence of a magnetic field, magnetic domains within certain crystals in the rock solidify with an orientation dictated by the applied field. This amounts to tiny magnetic fields embedded within the rocks. If the applied magnetic field changes as a succession of rocks cool, then there can be a record of the orientation of the earth's field as a function of the time of the cooling of the rocks. As we shall see, there is an explanation for this phenomenon within a model of recent creation and a worldwide Flood.

Beyond lacking a mechanism to rejuvenate a decayed magnetic field, there is at least one other problem for the conventional theory of multiple magnetic field reversals over billions of years. Each time the strength of the earth's magnetic field reaches zero, it may remain at zero (or be very weak) for a considerable time. The earth's magnetic field helps shield the earth from fast-moving charged particles from the sun and other astronomical sources. These charged particles are accelerated by the earth's magnetic field, deflecting them away and keeping them from striking the earth. Instead, the particles move in high orbits roughly around the earth's equator. Only near the earth's magnetic poles can the particles approach close to the earth. About 60 mi (100 km) up these particles collide with molecules in the air, mostly nitrogen and oxygen. These collisions cause ionization, stripping the nitrogen and oxygen atoms of their electrons. The electrons briefly move about freely until they are reabsorbed by other atoms in the earth's atmosphere. This reabsorption of electrons results in emission of light. We call this the aurora borealis, or, more colloquially, the northern lights (aurora australis or southern lights in the Southern Hemisphere). An aurora is a beautiful thing, but one can have too much of a good thing. If the earth had a weak or non-existent magnetic field, the solar wind would greatly increase the amount of radiation that we receive from cosmic sources. The earth's atmosphere would provide a great deal of protection from this onslaught, but at a price. The act of protecting the earth's surface would result in air being stripped away from the earth. In fact, this mechanism has been invoked to explain the sparse atmosphere of Mars. It appears that much of the earth's atmosphere would have been lost if the earth's magnetic field had been greatly reduced at various times during the earth's past.

Plate Tectonics

Our probing of the earth's interior has led to some interesting conclusions about the mantle. The *upper mantle*, down to about 400 mi (650 km) below the crust, is different from the *lower mantle*, which extends down to the core. The earth's crust is rigid and brittle. The topmost layer of the upper mantle is rigid and brittle too, so that layer and the crust can be treated as a unit. Geologists call this rigid crust and topmost upper mantle the *lithosphere*. The lithosphere varies in thickness, from about 50 mi (80 km) under the ocean basins to about 90 mi (150 km) under the continents (Figure 3.3). The upper mantle below the

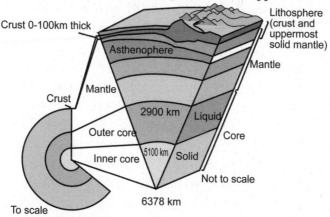

Figure 3.3. Contrast of lithosphere thickness under ocean basins versus under continents. (Wikimedia Commons).

lithosphere and extending down about 230 mi (370 km) is the *asthenosphere*. The composition of the asthenosphere is similar to the upper mantle above it. However, the higher temperature in the asthenosphere renders it plastic, meaning that its rock flows, though the flow is a bit sticky, like stiff modeling clay. Though the temperature generally increases with depth, pressure does too. The increased pressure in the bottommost part of the upper mantle causes the rock there to be stiffer, producing a transition layer between the asthenosphere and the lower mantle below. The mineral phase of the lower mantle is different from the upper mantle. Though the rock in the lower mantle flows, it flows much less than the upper mantle.

The interaction between the asthenosphere and the lithosphere has profound effects on the earth's surface. Forces within the earth cause the material in the asthenosphere to move. These flows are not in the same direction, but are in different directions at different locations on the earth. These flows drag the lithosphere atop the asthenosphere along with them. However, being rigid, the lithosphere does not flow as the asthenosphere does. Instead, the lithosphere is broken up into more than a dozen large pieces that we call *plates*. Geologists explain many of the features on the surface of the earth in terms of differences and interactions between these plates. This theory is called *plate tectonics*.

For instance, there are places in the asthenosphere where material upwells and moves outward along a line so that material moves away from the line on either side. On the earth's surface, the lithosphere is dragged in opposite directions on either side of the line, producing a rupture in the earth's surface. This rupture allows hot rock from deeper inside the earth to reach the surface. As the pressure on the hot rock decreases, some of this rock becomes molten. This rising material fuses to the crust on either side of the rupture producing new material on the earth's surface. This process occurs in the middle of the Atlantic Ocean along a roughly north-south line (Figure 3.4). The seafloor west of the line is moving westward, while the seafloor east of the line is moving eastward. This seafloor spreading slowly is increasing the width of the Atlantic Ocean. Along the line is a mid-ocean ridge, a series of mountain chains with a large valley, or rift, running along the spine. The Atlantic Ocean is so deep that very few of these mountains extend high enough to breach the surface and form islands.

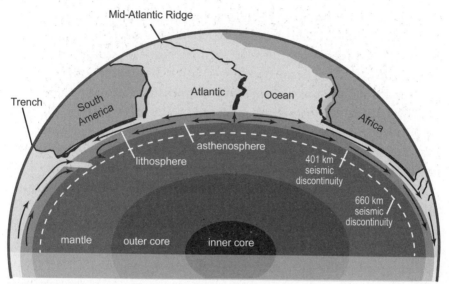

Figure 3.4. Mid-Atlantic rift zone (Ken Ham, ed., *The New Answers Book 1* [Green Forest, Arkansas: Master Books, 2006], 187).

There are several other places on the earth where seafloor spreading is adding new surface. Nearly all of these are in deep ocean. Why? Due to differentiation, the less dense material does not uniformly cover the earth's surface. As such, the majority of the less dense rocks on the earth are in the continents on the earth's surface. Indeed, the rock being added to the earth's surface via seafloor spreading comes from the earth's interior. Being from the interior, the added material is denser. Since the lithosphere is lying atop the plastic asthenosphere, the weight of the lithosphere is supported by the asthenosphere. The lithosphere effectively "floats" on top of the asthenosphere. The principle of hydrostatic equilibrium determines how high this material floats (Figure 3.5). Wherever

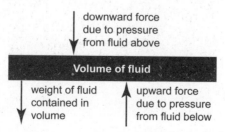

Figure 3.5. Illustration of the principle of hydrostatic equilibrium. (Wikimedia Commons.)

the weight of the lithosphere rocks is greater, they press down more deeply into the asthenosphere, but where the weight of lithosphere rocks is less, they press down more shallowly. Therefore, denser (heavier) lithosphere rocks are at lower elevation than less dense (lighter) lithosphere rocks. This is why the ocean basins are primarily made of denser rocks (oceanic crust), while the continents are made of less dense rock (continental crust). Some plates, such as the Pacific Plate, are made almost entirely of ocean crust. Other plates, such as the Arabian Plate, are mostly continental crust. However, most plates are like the North American Plate, made of both continental and oceanic crust.

The earth is not getting any larger, so if more material is added to the earth's surface in one place, material must be removed somewhere else. This often happens where an oceanic plate and a continental plate collide. An example of this is along the west coast of the Americas. Along much of the west coast of South America, the South American Plate is colliding with the Nazca Plate. As the plates collide, the Nazca Plate, being made of denser oceanic crust, bends downward, and passes under the South American Plate. This process is called *subduction* (Figure 3.6). The leading edge of the South American Plate

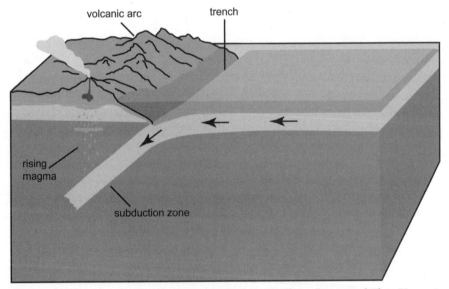

Figure 3.6. Illustration of subduction (Andrew A. Snelling, "How Did Plate Tectonics Get Started on the Earth." *Answers in Depth*. 2016. https://answersingenesis.org/geology/plate-tectonics/how-did-plate-tectonics-get-started/).

crumples, lifting part of the plate upwards, producing the Andes Mountains. As the Nazca Plate descends, increased temperature causes the upper mantle material above it to melt. This molten material (magma) rises through cracks and faults into the continental crust above. If any of the magma makes it to the surface, we call the molten rock lava, which is what we see in a volcanic eruption. It is no accident that volcanoes are often found in mountain-building regions where oceanic and continental plates collide, such as along the west coast of the Americas.

The Pacific Ocean is ringed by similar plate interactions. The western coast of Central America is a collision of the Cocos Plate with the North American and Caribbean Plates. Most of the western coast of North America is a collision between the North American Plate and the Pacific Plate, with the smaller Juan de Fuca Plate sandwiched in between. In the western Pacific, the Pacific Plate collides with several other plates, forming several island chains, such as Japan, the Marianas Islands, and the Philippines. There are many volcanoes along these chains, which is why this sometimes is called "the ring of fire."

Plates do not always collide head-on. Sometimes plates have a component of motion parallel to their line of collision. This causes the plates to grind past one another along fault lines. The sliding motion is not constant because friction binds the plates into place. Over time, forces build along the line of contact. When the forces grow great enough to overcome the friction locking them into place, an earthquake results. The motion generally is along faults, many of which can be traced on the surface. Mountain-building regions, such as along the West Coast of the United States, can have many faults (e.g., the San Andreas Fault).

Sometimes continental plates collide, such as in northern India, where the Indian and Eurasian Plates collide. Since neither plate is oceanic, subduction does not happen so readily. However, often the leading edge of one plate pushes under the leading edge of the other plate, and both plates crumple and uplift. This leads to very high and broad mountain ranges. This is why the Himalaya Mountains are here, and the Tibetan plateau beyond.

Plate tectonics is a robust theory that explains much of the earth's surface. However, there are a few odd things that the theory does not explain directly. For instance, there are some regions of volcanic activity far from plate boundaries. Examples are the Hawaiian Islands and the Yellowstone hotspot.

Geologists explain these by a region of hot rock called plumes rising from deep beneath the crust, but in one region, not along a line of upwelling rock, as is happening on mid-oceanic ridges. There are some deep faults nowhere near plate boundaries, such as the New Madrid Fault Line near New Madrid, Missouri. Faults such as these may be remnants of an earlier episode of plate tectonics, but there is much about such things that is not yet understood.

Plate tectonics generally is interpreted in terms of an earth that is billions of years old. In this view, the plates move slowly. Also within this view is the idea that in the past the continental plates were combined into a single supercontinent several times. Each supercontinent broke up, only to reform via collisions, and then to break up again. Geologists call the last supercontinent Pangaea. Most geologists think that Pangaea formed about 300 million years ago and broke up about 175 million years ago. The supercontinent prior to Pangea was Rodinia. By conventional dating, Rodinia formed about a billion years ago, and it broke up at least 600 million years ago.

Given that discussion of plate tectonics usually is within the context of billions of years, this comes across as evolutionary, and so biblical creationists, who believe that the world is only thousands of years old, understandably could be suspicious of plate tectonics. How have biblical creationists responded? Some creationists have rejected plate tectonics and offered their own interpretations of the earth's surface with limited success. However, a very robust theory has emerged that embraces much of plate tectonics, albeit on a much shorter timescale. Catastrophic plate tectonics (CPT) accepts that the earth's surface today consists of plates that can and have moved on the earth's surface. Furthermore, CPT agrees that there were two supercontinents in the past, but no more than two. Any idea that there were other supercontinents earlier is conjecture at best.[1]

---

[1] For more details about the CPT model, see Steven A. Austin et al., "Catastrophic Plate Tectonics: A Global Flood Model of Earth History," in *Proceedings of the Third International Conference on Creationism*, edited by R. E. Walsh (Pittsburgh, Pennsylvania: Creation Science Fellowship, 1994), 609–612, available at https://answersingenesis.org/geology/plate-tectonics/catastrophic-plate-tectonics-global-flood-model-of-earth-history/. See also John R. Baumgardner, "Catastrophic Plate Tectonics: The Geophysical Context of the Genesis Flood," *TJ* 16, no. 1: 58–63, available at https://answersingenesis.org/geology/plate-tectonics/catastrophic-plate-tectonics-geophysical-context-of-genesis-flood/); and Andrew A. Snelling, *Earth's Catastrophic Past: Geology, Creation and the Flood* (Dallas, Texas: Institute for Creation Research, 2009), 683–706.

In the CPT theory, the earth as it was initially created had one supercontinent (Rodinia). The cold, pre-Flood oceanic crust was denser than the warm, less dense upper mantle. It is possible for denser material to float atop less dense material, but this is an unstable situation. At the time of the Flood, this instability led to a topple, with denser surface material rapidly subducting down into the earth's mantle toward its bottom. This downward motion displaced mantle material in the paths of the descending plates, leading to an upwelling of mantle material elsewhere. The tectonic upheaval raised ocean basins, dumping water onto Rodinia. Rifting of the crust permitted water under pressure in the mantle to escape in superheated steam jets that rose at supersonic speed. Most of the resulting rain was from ocean water carried aloft by the jets. This process of runaway convection within the earth was far faster than anything observed today. The rapid convection dragged the lithosphere in different directions, leading to the catastrophic break-up of Rodinia. The pieces of Rodinia sped around the globe, crashed together to form Pangaea briefly before splitting up again. Some portions of these plates continued to collide, but no more supercontinents formed. Once the denser material had sunk to the bottom of the earth's mantle, the runaway convection ceased, and the catastrophic motion of plates slowed to the much more gradual plate motions that exist today.

The exact details and timeline of the CPT processes are still being worked out. One preliminary result is that the rapid chaotic motion produced a rapidly changing magnetic field. There could have been several episodes of magnetic field reversals during this process, before settling into the reasonably stable, yet exponentially decaying field that exists today.[2] This process would have worked out in a few months, so the times when the earth's magnetic field was near zero would have been very brief. Therefore, there was no serious harm to the earth's atmosphere during these short intervals. The CPT model is a work in progress, with refinements surely to come.

---

[2] The mechanism here is very different from the gradual evolutionary model with an unidentified dynamo regenerating a new magnetic field once the preceding magnetic field disappeared.

Earth's Atmosphere

The earth's atmosphere mostly is nitrogen (78%) and oxygen (21%). Notice that nitrogen and oxygen together comprise 99% of the earth's atmosphere. The remaining 1% is a mixture of various gases. The nitrogen and most of the oxygen in the earth's atmosphere are *diatomic*, meaning that their molecules consist of two atoms (in standard chemical notation, we write them as N_2 and O_2). This is very significant. There are two other kinds of gases: monatomic and polyatomic. The molecules of monatomic gases consist of single atoms. These are the noble gases, such as helium (He), neon (Ne), and argon (Ar). Argon comprises most of the remaining 1% of the earth's atmosphere. Polyatomic gases have three or more atoms in their molecules. Examples of polyatomic gases are carbon dioxide (CO_2), ozone (O_3), and water vapor (H_2O).

Polyatomic gases block infrared (IR) radiation. IR radiation sometimes is called *heat radiation*, because warm objects cool by emitting in the IR part of the spectrum. This is how a greenhouse works. The glass of a greenhouse allows visible sunlight to pass through so that plants can use photosynthesis. The sun's radiation also warms objects in the greenhouse. As these objects warm, they radiate in the IR, but most glass is opaque to IR radiation. Therefore, the IR radiation is reflected off the glass and back into the greenhouse. Objects in the greenhouse reabsorb the energy, further driving up their temperature. As the objects warm, conduction transfers some of their heat to the air, warming the air of the greenhouse. This is why greenhouses are so warm on sunny days.

Since polyatomic gases block IR radiation, we sometimes call them *greenhouse gases*. They act to trap heat in a planet's atmosphere. The earth's atmosphere has enough greenhouse gases to maintain a warmer temperature than it would if the greenhouse gases were not present. This is particularly important overnight. If there were no polyatomic gases in the earth's atmosphere, the temperature at night would plunge far below freezing, even in the tropics. Therefore, some polyatomic gases are important to have in the atmosphere. On the other hand, one can have too much of a good thing. As we shall see, Venus has far too much greenhouse gases in its atmosphere, which causes an extreme greenhouse effect, resulting in a very high surface temperature.

You might think that having more oxygen in the atmosphere would be better, but it is not. Most organisms do not work as well in high-oxygen environments. Furthermore, combustion is much more violent in air that has a high oxygen content. This is why warning signs are posted near tanks that contain oxygen. It turns out that the composition of the earth's atmosphere is ideal for life. Even evolutionists are convinced of this. Those who believe that there must be many places in the universe where life exists think that searching for planets with atmospheres similar to earth's offers the best hope of finding life elsewhere.

The Canopy Theory

As mentioned in Chapter 1, at one time many biblical creationists believed the canopy theory. The canopy theory proposes that the *rāqîaʿ* that God created on Day Two is the earth's atmosphere. In this view, the waters above the *rāqîaʿ* formed a canopy, or layer, above the earth's atmosphere. The canopy collapsed at the time of the Flood to provide a huge amount of rainfall as one of the two sources of the Flood (cf. Genesis 7:11). Prior to its collapse, the water canopy supposedly explained several things in the book of Genesis. For instance, the water canopy supposedly provided a greenhouse effect that kept the earth warm and wet. With it, the earth functioned like a terrarium. If so, this would explain the meaning of Genesis 8:22, which reads,

> While the earth remains, seedtime and harvest, cold and heat, summer and winter, day and night, shall not cease.

According to the canopy theory, the worldwide warm weather prior to the Flood had no seasons. Genesis 2:5–6 indicates that initially there was no rain on the earth, but that instead water issued up from the ground to water the earth. Rain is not mentioned again until the Flood (Genesis 7:4), which occurred more than 1600 years after Creation. Presumably, the text's failure to mention rain between these two verses suggests that the lack of rain was not just the *initial* condition on the earth, but was the *ongoing* climate regime until the Flood. If there were no rain before the Flood, then there could not be a rainbow prior to the Flood. This would explain the sign of God's covenant with Noah (Genesis 9:12–17), in that this supposedly was when rainbows first became possible. Finally, the canopy purportedly protected the earth's surface from harmful radiation, leading to longevity among the antediluvians. Lifetimes rapidly decreased after the Flood until shortly after the time of Abraham, when lifespans approached those of today.

Given the many benefits of the canopy theory, why did creation scientists abandon the model in recent years? There are several factors. First, the canopy model was based upon a questionable understanding of certain biblical passages. As shown in Chapter 1, and in Chapter 2 of *The Created Cosmos: What the Bible Reveals about Astronomy*, the *rāqîaʿ* is better understood as what we today would call outer space and at least part of the atmosphere rather than just the atmosphere. Or consider Psalm 148:4, which reads:

> Praise him, you highest heavens,
> and you waters above the heavens!

What are these waters above the heavens? This sounds like the waters above the *rāqîaʿ*. This is the only other time in Scripture where waters aloft are mentioned. While the word *rāqîaʿ* is not used here, the *rāqîaʿ* is equated with heaven in Genesis 1:8 in the context of the Day Two creation account (Genesis 1:6–8), strongly implying that this is the correct identification of these waters in Psalm 148:4. Psalm 148 clearly was written long after the Flood, yet it mentions the waters above as if they were still there, but this would not be possible if those waters had fallen to earth during the Flood.

While it is a common belief that there was no rain before the Flood, Genesis 2:5–6 does not clearly state this. Rather, it may indicate only that it did not rain during the Creation Week, but rained soon thereafter. It is a questionable interpretation of Genesis 2:5–6 that leads to the conclusion that there was no rain before the Flood. The same is true concerning the rainbow. It is not necessary that rainbows did not exist prior to the Flood for the rainbow to be the sign of God's covenant with Noah. God could have assigned new meaning to something that already existed, much like Christ assigned new significance to the bread and the cup from the Passover meal when He inaugurated the New Covenant. It is reading into the text to insist that rainbows did not exist before the Flood. Furthermore, the promise of Genesis 8:22 does not indicate that there were no seasons prior to the Flood. The same verse also mentions day and night, but those clearly existed before the Flood, so it again is reading into the text to suggest seasons did not exist before the Flood. Finally, there are ways that longevity before the Flood and gradual decline in lifetimes afterward could be explained other than by invoking a protective canopy.

Additionally, there are physical problems with the canopy theory. Creation scientists attempted for years to develop a physical model of how the canopy might have worked, but they failed. Problems include the lack of a mechanism to keep the water aloft. If supported by the earth's atmosphere, hydrostatic equilibrium would have added significantly to atmospheric pressure then. Also the water probably would have scattered or absorbed starlight, making it difficult for the stars to be seen from the earth's surface. The stars could not have fulfilled their purposes if they were not visible. Thus, the canopy theory is an idea whose time has gone.

How Old Is the Earth?

Most scientists believe that the earth (and the rest of the solar system) is almost 4.6 billion years old. The six-day creation account of Genesis, along with the genealogies of Genesis 5 and 11, and chronological information given elsewhere in Genesis and the Old Testament indicate that the world is a little more than 6000 years old. Obviously, there is great disparity in these two ages. There are numerous ways that Christians have attempted to reconcile this disagreement. One attempt is the gap theory, the classic version of which states that there was a vast amount of time between Genesis 1:1 and Genesis 1:2. Another approach is the day-age theory, the belief that each of the six days of the Creation Week were long periods of time. These two approaches attempt to hold on to some concept of the Creation Week describing historical events. Other Christians minimize the details of the creation account in favor of some more important, broader meaning. This deeper meaning might be the fact that God created us, even though the actual details of how He did so are unimportant. The framework hypothesis is one such approach, proposing that the structure of the creation account is thematically (rather than chronologically) arranged, and that the temporal aspect of the narrative is not reflective of actual time. Alternatively, some see the creation narrative merely as a theological polemic against ancient Israel's pagan neighbors.[3] This view argues that the creation account was meant only to refute the beliefs of the pagan nations that surrounded Israel and was not intended to convey a literal record of history.

---

[3] There do exist in the creation narrative many polemical elements, but it does not follow that the account is rendered ahistorical due to the presence of a polemic. For more information, see Chapter 1.

It is not the purpose here to refute these ideas, for this has been done elsewhere.[4] The author's position is that all of these attempts to accommodate the modern idea that the creation is billions of years old are flawed. While giving lip service to the authority of Scripture but bowing to the supposed wisdom of modern science, it is clear that supporters of these various views really judge the statements of the biblical text by whatever modern science supposedly has established—that the earth is billions of years old. This is backwards—the Bible ought to sit in judgment over the interpretations of data in modern science, not the other way around. If the world is only thousands of years old, might there be evidence of that? Yes, and some of those evidences will be discussed in this book (for instance, see the discussion of the earth's magnetic field earlier in this chapter).

An important question is why so many scientists think that the world is billions of years old. Most people seem to think that it has to do with evidence, that scientists set out objectively to determine the earth's age, apart from any bias in the matter, and simply reached this conclusion. In reality, nothing could be further from the truth, because the conclusion is driven by the assumptions made. The major assumption is that evolution is the mechanism by which the world came to be. We don't see evolution occurring before our eyes. This certainly is the case with biological evolution. We see variation of traits within groups of organisms, but we don't see one kind of organism transforming into a totally new kind of organism. The same is true of geological evolution. For instance, we don't see mountains building before our eyes, yet most geologists think that mountains formed by collisions of plates and uplift. If evolution is the correct explanation of how our world came to be, then it must proceed very slowly for us not to observe it directly. And this then requires that the world be very old. How old must it be? It depends upon the rate of evolution that one assumes. With the *assumption* of the rate of evolution, one can then estimate how old the world must be.

This is illustrated by the generally assumed age of the earth in the latter nineteenth century. By then, both biologists and geologists largely had become convinced that evolution was true. Both biologists and geologists were convinced that the length of time required for evolution to account for what they saw in the world was 100 million years, so this was the age of the earth

---

[4] For example, Terry Mortenson and Thane H. Ury, editors, *Coming to Grips with Genesis: Biblical Authority and the Age of the Earth* (Green Forest, Arkansas: Master Books, 2008).

established at that time. However, what did the data say? Lord Kelvin, one of the most significant physicists of the nineteenth century, used the best science of his day to estimate an upper limit to the age of the earth and the age of the sun. In the case of the earth, Kelvin assumed that the earth began as a hot mass and had slowly cooled, which was the concept of the earth's origin at the time. For the sun, Kelvin assumed that the sun was powered by gravitational contraction, which also was the preferred mechanism for the sun's generation of energy at the time (see Chapter 6). Both methods converged on an upper limit of the age at 30 million years, scarcely one-third that age assumed by evolutionists. Notice that both of these dates (consistent with one another) were upper limits—the actual age of the earth and sun could be far less, but they could be no more. Did this dissuade many scientists from believing that the earth was 100 million years old? No. To abandon such a long age would force the abandonment of evolution, which was unthinkable to most scientists. That is, they believed in an old age for the earth in spite of, not because of, the best evidence available.

Some would quickly protest that today people believe that the earth is far older than 100 million years. Furthermore, the basis of Kelvin's calculations has been supplanted by the discovery of radioactivity at the end of the nineteenth century. The radioactive elements in the earth's interior release heat, greatly increasing the time that the earth's interior can remain hot. Furthermore, we now think that the sun is powered by nuclear reactions in its core, greatly expanding its lifetime too. However, this entirely misses the point. Scientists of more than a century ago persisted in a belief that contradicted what the best science then available told them. That is, data was not important. The situation has not changed much today. Belief in vast age came first, followed by quantitative methods that "confirmed" the assumption. Along the way, scientists tended to ignore data that suggested ages that were significantly different.

Today, most people who believe that the earth is almost 4.6 billion years old claim radiometric or radioactive dating as the evidence of this. How does radioactive dating work? Particular *isotopes*[5] of some elements are radioactive.

---

[5] All atoms of a particular element have the same number of protons, but isotopes have different numbers of neutrons in their nuclei. For instance, there are three isotopes of carbon that exist on earth, C-12, C-13, C-14. The numbers represent the atomic weights, the sums of the number of protons and neutrons in each isotope. All carbon atoms have six protons. C-12 has six neutrons, C-13 has seven neutrons, and C-14 has eight neutrons. C-12 and C-13 are stable, but C-14 is radioactive.

As atoms of a radioactive isotope decay, they form an isotope of another element. For instance, C-14 decays into N-14.[6] We call the original radioactive isotope the *parent*, and we call the isotope that the parent decays into the *daughter*. The length of time required for half of the parent to decay into the daughter is the *half-life*. Notice that decay is geometric. That is, half of the parent will decay in each half-life. Therefore, after one half-life, only one-half of the parent will remain; after two half-lives, only one-half of one-half, or one-quarter, of the parent will remain. After three half-lives, one-eighth of the parent will remain. One-sixteenth of the parent will remain after four half-lives, and so forth. Many igneous rocks include radioactive isotopes when they form. These radioactive isotopes generally reside in small crystals within the rock. Assuming that no daughter element is included within the crystals, the ratio of the amount of daughter and parent materials ought to reveal the number of half-lives since the rock formed. Radioactive dates *generally* are much older than thousands of years, far older than the biblical timeline would permit. This methodology sounds straightforward enough, so how does one who takes the biblical timeline seriously respond to this?

First, one must realize that there are some restrictions on what radiometric dating actually can accomplish. Radiocarbon dating, or C-14 dating, appears to be the form of radiometric dating most familiar to the general public. A common misconception is that radiocarbon dating can be used to find the earth's age. However, C-14 has a relatively short half-life (5730 years). After ten half-lives or so, it becomes difficult to measure the meager amount of parent isotope remaining, so radiocarbon dating can be used only to date things to no more than 50,000–100,000 years old. Therefore, radiocarbon dating cannot be used to establish an age for the earth that is billions of years. Furthermore, unlike other radiometric dating methods, radiocarbon dating is not used on rocks. Rather radiocarbon dating is used on only preserved organic materials, such as bone, skin, fur, and plant fibers.

Another common misconception is that radiometric dating can be used to determine the ages of fossils. Radioactive dating of rocks generally works on igneous or metamorphic rocks, but rarely on sedimentary rocks. Fossils generally are found in sedimentary rocks, and thus cannot be dated using

---

[6] N-14 is an isotope of nitrogen having seven protons and seven neutrons in each atom.

radiometric dating. For instance, the Grand Canyon exposes sedimentary layers of rock about a mile thick. Radiometric dating cannot be used to find the ages of any of those layers. There are many places at the bottom of Grand Canyon where there are igneous rocks, so one can obtain radiometric dates for them. Furthermore, there are a few places in Grand Canyon where there are igneous rocks on top of rock layers making up Grand Canyon, so radiometric dating can be used to obtain ages for them. However, at best this brackets the ages of all the rocks in between. Similar circumstances and reasoning exists for bracketing the ages of sedimentary rocks elsewhere on earth.

There are at least three assumptions that one must make in using radiometric dating. The first assumption is that none of the daughter isotopes were incorporated into the crystals making up the rock along with the parent isotopes. Igneous rocks form from a melt, so there is no reason why none of the daughter was not originally present in the melt. On the other hand, chemical differences between the parent and daughter elements could prevent inclusion of daughter isotopes along with the parent isotopes. The method of isochron dating supposedly eliminates the problem of primordial daughter isotopes contaminating the samples. The second assumption is that we know the value of the half-life, and that the half-life has remained constant. Half-lives of various radioactive isotopes usually are measured under controlled laboratory conditions, so it is likely that we do know them as they exist today. However, has the decay rate remained constant throughout time? The third assumption is that there has been no leaching of either the parent or the daughter isotopes into or out of the rock being sampled. Because the parent and daughter isotopes are different elements having different chemical properties, it is quite possible that one or the other has selectively been removed or added to the samples. If the daughter isotopes have been added or the parent isotopes have been removed, the radioactive date will be older than the actual date.

There are numerous examples of radiometric dating having given false dates. Consider the bracketing of sedimentary rocks as described above with sections of Grand Canyon. It would seem to be axiomatic that the lower igneous rock ought to be older than the upper igneous rock. However, two studies have together shown that the basalt lavas atop Grand Canyon are 1.1 billion years old, while the basalt lavas within the basement rocks, underneath all the other rocks, are 1.0 billion years old by the same radiometric dating method.[7] Obviously, the older rocks cannot be on the top. There are examples of rocks of known ages (because they are the result of observed volcanic activity) that have radiometric ages far older than the known ages. How do those committed to billions of years explain these documented discrepancies? Generally, these discrepant ages are dismissed as due to contamination or to improper use of the radiometric dating method employed. However, these appeals are made only when a radiometric date clearly is in error. There are many other radiometric dates that are omitted, but not because they have contradictory bracketing or because they contradict observed ages. Rather, some dates are thrown out, simply because they do not conform to the ages assumed by the evolutionary model. If radiometric dating were such a straightforward, exact science, there would be no need to omit so many so-called discordant ages.

Still, radiometric dating generally produces ages that are far older than a few thousand years. This is because most samples contain significant amounts of daughter products along with their parent isotopes, suggesting that considerable radioactive decay has occurred. At the current rate of decay, that would suggest that the rocks involved are very old. How do biblical creationists respond to this? One possibility is to question whether there was complete segregation of the parent and daughter isotopes in the melt from which the igneous rocks being dated formed. If some daughter product is included, then this would give the appearance of great age in a relatively young rock. We know that this definitely occurs, as already described above. Another possibility is that leaching has preferentially removed parent isotopes and/or added daughter products. Still another possibility is to question whether radioactive decay has been constant in the past. This was the approach of a research initiative known as *Radioisotopes and the Age of the Earth* (RATE), a joint project of the Institute for Creation Research and the Creation Research Society.[8] This study presented evidence that there was a brief period of rapid radioactive decay, perhaps at the time of the Flood.

---

[7] See Andrew A. Snelling, "The Fallacies of Radioactive Dating of Rocks: Basalt Lava Flows in Grand Canyon," *Answers Research Journal* 1: 66–69, 2006, available at https://answersingenesis. org/geology/radiometric-dating/the-fallacies-of-radioactive-dating-of-rocks/.

[8] For more information, see Larry Vardiman, Andrew A. Snelling, and Eugene Chaffin, editors, *Radioisotopes and the Age of the Earth*, vols. 1 and 2 (El Cajon, California: Institute for Creation Research, and Chino Valley, Arizona: Creation Research Society, 2000 and 2005); and Don DeYoung, *Thousands, Not Billions: Challenging an Icon of Evolution, Questioning the Age of the Earth* (Green Forest, Arkansas: Master Books, 2005).

Even if one were to accept the reliability of radiometric dating, the age of the earth still is a murky subject. The earth is a geologically active body. Extrapolating this geologic activity over billions of years, it is not likely that many rocks survived on the earth's surface in their primordial condition. Hence, the radiometric dates of all terrestrial rocks ought to be younger than the actual age of the earth. Most geologists think that the Canadian Shield has some of the oldest extant rocks on earth. The oldest generally undisputed rocks from the Canadian Shield have a radiometric date of 4.03 billion years, more than a half-billion years short of the assumed 4.55-billion-year age of the earth. Some small zircon crystals contained in rocks in the Jack Hills region of Western Australia have been dated to 4.40 billion years. These small crystals are presumed to be fragments of a much earlier rock that no longer exists. Supposedly, the surviving zircons were, by means of geological processes, incorporated into a much younger rock. The moon is much less geologically active, so it was thought for a long time that the moon might yield even older rocks. One rock brought back to earth by Apollo 16 has a radiometric date of 4.46 billion years. The oldest radiometric dates claimed are for certain meteorites. Many meteorites contain chondrules, small round particles that melted and cooled in space. Some chondrules have radiometric dates slightly older than 4.55 billion years. Chondrules are thought to be samples of the original material that formed the solar system, so their radiometric dates are thought to reflect the age of the solar system and hence the earth.

The 4.55-billion-year age of the earth depends upon several assumptions unrelated to the radiometric dates. Buried deeply in this is an assumption of an evolutionary history of the solar system. The solar system supposedly formed from a cloud of gas and dust. Much of the matter in the cloud collapsed to its center to form the sun. The remaining material flattened into a disk, in which small particles, planetesimals, began to form. It is not known how these particles began to stick together. Some planetesimals became large enough that their gravity could accumulate additional planetesimals. These larger bodies eventually grew into the planets, satellites, asteroids, and comets that we see in the solar system today. The different compositions in these bodies supposedly are the result of where they formed in the solar system. It is within this general framework that radiometric dates are interpreted. These assumptions are contrary to the biblical account of creation, so we know that this scenario is untrue. This frees creation scientists to interpret all the data, including the radiometric dates, with a different paradigm, leading to very different conclusions.

CHAPTER 4

The Lesser Light to Rule the Night: The Moon, Our Nearest Neighbor

The earth has one natural satellite, the moon. Most of the other planets have natural satellites too.[1] Notice that the proper term is *natural satellite*. After the invention of the telescope four centuries ago, astronomers discovered some of these other natural satellites. Since they resembled the earth's moon in that they orbited planets, it became common to refer to these satellites as *moons*. While this remains a common practice, the moon is the proper name of the earth's natural satellite. To maintain that distinction, astronomers generally refrain from calling the natural satellites of the other planets *moons*, especially in more technical contexts. We will observe that practice here.

The moon's diameter is about one-quarter that of the earth's and, correspondingly, the moon's volume is about $1/_{64}$ that of the earth. The moon's density is a little more than 3.3 gm/cc, compared to the earth's 5.5 gm/cc. To account for this difference in density, the moon must lack much of the denser material that the earth has. We saw in the previous chapter that the earth contains much iron and nickel in its core. Therefore, it is most likely that the moon is deficient in those two elements. Indeed, one can account for the overall density of the moon if the moon were made entirely of rock that is not that different from some of tahe moon's surface rocks. However, most planetary scientists think that the moon has undergone extensive differentiation so that the moon has at least a small core.

---

[1] Mercury and Venus are the only exceptions.

The Moon's General Appearance

The moon has no light of its own. It merely reflects the light of the sun. The moon orbits the earth once a month. As discussed in Chapter 2, the moon's orbit is the basis of the month, and the word *month* even comes from the Anglo-Saxon word for the moon. As the moon orbits the earth, we see varying amounts of the lit half of the moon. We call the changing amount of the lit half of the moon that we see the lunar *phases*. Once per month, the moon is opposite the sun in the sky. This phase is called full moon. We call this phase full moon because the moon appears fully illuminated. Halfway between two full moons, the moon is between the earth and sun. At this point, the lit half of the moon faces away from the earth, so the moon presents its dark side to the earth. The dark half of the moon is not very bright, which would make it difficult to see under any circumstances. However, because the moon appears so close to the sun at this time, the sun's overwhelming brightness makes it impossible to see the moon for two or three days. We call this phase new moon, because we usually consider this to be the start of the cycle of lunar phases (Figure 4.1).

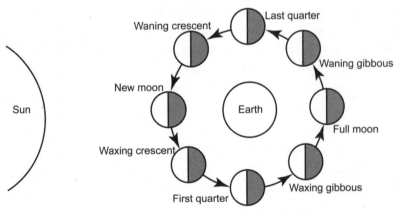

Figure 4.1. Phases of the moon.

A few days after new moon, the moon becomes visible in the western sky after sunset. Most of the moon's lit half is facing away from us so that only a thin sliver of the lit half is visible. We call this phase the crescent phase. More specifically, it is the waxing crescent, because each evening the amount of the lit half that we see increases, so the thickness of the crescent grows, or waxes, from night to night. About a week after new moon, we see half of the lit half of the moon, so the moon appears half lit. We call this phase first quarter. Notice that we call this phase a quarter, even though the moon appears half lit. Why? The *quarter* here does not refer to the moon's appearance, but instead refers to the distance that the moon has traveled thus far, one-quarter of its orbit since new moon. A few days after first quarter, we see most of the moon's lit half, with just a sliver, or crescent, of the unlit side of the moon showing. We call this a gibbous phase or, more specifically, the waxing gibbous, because the lit portion that we see continues to increase from night to night. About a week after the first quarter, or two weeks after the new moon, the moon reaches the full phase.

After the full moon, the lunar phases progress in the opposite direction. Instead of the lit portion increasing each night, the lit portion shrinks. We call these the waning phases, rather than the waxing phases. The full moon is followed by waning gibbous, then third quarter, waning crescent, and then finally new moon, which restarts the cycle. Third quarter is about a week after full moon, and new moon is about a week after third quarter, or two weeks after full moon. In the Northern Hemisphere, the waxing phases are lit on the right side of the moon, while the waning phases are lit on the left side of the moon. This is reversed in the Southern Hemisphere. All waxing phases are visible early in the evening, while all waning phases are visible in the early morning, shortly before sunrise.

The moon exhibits synchronous rotation, which means that the moon rotates on its axis at the same rate that it orbits the earth. The upshot is that the moon always keeps one face toward the earth. Consequently, we only see about half the moon's total surface.[2] Until the space age, we had no idea what the far side of the moon looked like. Of course, now the far side of the moon has been mapped with the same detail as the near side of the moon. There is much confusion about the term *dark side of the moon*. In popular culture, this seems to refer to a side of the moon that is perpetually in darkness, indicating somewhere remote and forbidding. However, as the moon rotates once per month, all of the moon's surface is illuminated by the sun,[3] so there is no perpetually dark

---

[2] We see exactly half of the moon's total surface at any given time. However, slight irregularities, called librations, allow us to peek around the edges of the moon. Over time, we can manage to see nearly 59% of the moon's surface.

[3] However, there are a few spots on the floors of deep craters near the moon's north and south poles where sunlight never reaches.

side of the moon. Until the back side of the moon was photographed in 1959, astronomers sometimes referred to the far side of the moon as the dark side of the moon, but the term *dark* referred to the unknown. Similarly, in the early eighteenth century, Africa came to be called *the dark continent* because little was known of its interior. That terminology eventually disappeared as much of Africa was explored and mapped by Europeans.

Why does the moon have synchronous rotation? Many natural satellites of the solar system rotate synchronously, so it is a common characteristic of satellites. Because planets are far more massive than their satellites and satellites orbit so closely to their planets, the planets' gravity exerts large tidal forces on their satellites. These strong tides stretch satellites from their normally spherical shapes into slightly oblong shapes (Figure 4.2). If a satellite is not rotating synchronously already, then the satellite's rotation will carry this bulge ahead of the line connecting the satellite to the planet (or it could lag behind, if the satellite is rotating very slowly). The planet's gravity will pull on this bulge, but it pulls more on the nearside of the bulge than on the satellite's far side. This difference in force produces a torque, which slows the rotation of the satellite, eventually bringing the bulge in line (or speeds the satellite's rotation, if it was initially rotating too slowly). Eventually, this achieves synchronous rotation. The time required to do this generally is acknowledged to take much more than a few thousand years, so why do the moon and so many other natural

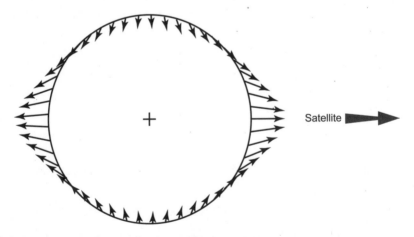

Figure 4.2. Illustration of tidal forces (Wikimedia Commons).

satellites rotate synchronously? Biblical creationists could argue that God created the moon and many other natural satellites with synchronous rotation. However, why would He do that? The most likely answer is that satellites were designed that way to fulfill some purpose, though we do not yet know what that purpose might be.

Lunar Surface Features

Everyone is familiar with the many craters on the lunar surface.[4] However, these cannot be seen without a telescope, so lunar craters were not known until four centuries ago, after the invention of the telescope. What causes craters? Historically, there have been two schools of thought—the impact theory and the volcanic theory. The impact theory holds that many objects, called *meteoroids*, have struck the moon's surface. A meteoroid is any small body orbiting the sun that has the potential to collide with a planet or satellite. Technically, any object more than a few meters across is an asteroid, but for now, I will use the term to refer to any impacting body. Meteoroids have high velocity, typically on the order of a few tens of kilometers per second. This corresponds to a tremendous amount of kinetic energy. When a meteoroid strikes the lunar surface, it rapidly comes to a stop. By the principle of conservation of energy, the kinetic energy must transfer to the lunar surface. This energy is manifested as an explosion that excavates a crater. The diameter of an impact crater greatly exceeds the size of the impacting body.

The volcanic theory holds that volcanic events on the moon produced craters. There are numerous examples of volcanic craters on the earth's surface, but until relatively recently no impact craters were recognized on earth. Therefore, for much of the past four centuries the volcanic theory was dominant. However, by the middle of the twentieth century, that assessment began to change. Today, we believe that nearly all the noticeable craters on the moon resulted from impacts. There are some volcanic craters on the moon, but they are relatively small.

Even to the naked eye, the moon appears to have two distinct types of terrain. This is particularly noticeable near a full moon, when one can readily

---

[4] We get the word *crater* from the Latin word for bowl, which is very descriptive of the geologic structure.

see darker and lighter regions on the lunar surface. Through a telescope, the two types of regions appear even more different. The darker regions are relatively smooth, with few craters, while the lighter colored areas are heavily cratered. When astronomers began to study the moon through telescopes four centuries ago, they thought that the darker, smoother areas might be bodies of water, so they called the darker regions *maria* (pronounced *mǎ'-rēǎ*), which is Latin for seas. Note that the word *maria* is plural; the singular (for sea) is *mare* (pronounced *mǎ'-rā*). The lighter regions are the lunar highlands, so called, because they are at higher elevation than the maria. Why are the lunar maria and highlands so different in brightness? It is because they are made of different material. The maria are made of rocks that resemble basalt on the earth. On the other hand, the highlands consist of rocks that are similar to terrestrial granite. On earth, basalt generally is a very dark rock, while granite tends to be much lighter. This difference in composition was inferred long before we had direct rock samples gathered by Apollo astronauts in the late 1960s and early 1970s.

Why do the maria have so few craters, while the highlands have many craters? One possibility is that most impacts just happened on the highlands but somehow missed the maria, but that would have been highly improbable. A better explanation is that all the moon was subject to heavy cratering, but that most of the craters on the maria have been erased. A clue to piecing together this puzzle is that some of the smaller maria clearly are circular shaped, looking a bit like very large craters. Even the large maria that are not circular look as if they are comprised of a series of overlapping circles. Indeed, astronomers agree that the maria resulted from very large impacts. The craters excavated were so large (nearly ten times larger than the largest normal craters) that astronomers refer to them as impact basins rather than craters. The impacts that formed the impact basins required very large meteoroids, or, more properly, asteroids. These impacts were so violent that they fractured the lunar surface down to the depth where there was molten material. The fractures acted as conduits for magma to move upward and erupt onto the lunar surface as lava flows. The lava flows filled the impact basins, and even spilled into nearby low-lying regions in some cases. The large impacts that formed the impact basins would have obliterated any craters already present, and the volcanic overflow would have provided the smooth, dark surface that we see on the lunar maria now.

To recap, the initial lunar surface probably was granite and was heavily cratered by impacts. Then there was a series of very large impacts that formed some of the largest craters yet. These craters are so large that they warrant a special name, impact basins. The deep fractures caused by these large, late impacts permitted molten material (basalt) from deep in the lunar interior to flood the impact basins. Once the volcanic overflow cooled and hardened, further impacts peppered the relatively smooth maria.

This scenario explains many things. It explains why the maria generally are circular or are overlapped circular features. It explains the color difference between the maria and highlands. It explains why the highlands have so many craters, while the maria do not. The highlands bear testament to all the impacts that they have endured (though some probably were obliterated by later impacts), but the maria only record the impacts that occurred after the volcanic overflow had hardened. It also explains why the highlands and maria have different elevations. Basalt is a denser rock than granite, so hydrostatic equilibrium has forced the maria to lower elevation while the highlands have ridden higher up. A similar thing is present on the earth, where there is a difference between the density of the rocks that make up the continents versus those that make up the ocean depths. Being less dense, the continental crust is higher, while the oceanic crust, being denser, lies at lower elevation.

Notice that this scenario also amounts to a history of the lunar surface. Indeed, we can infer some history of the moon, and we can even determine relative ages for some features. For instance, the craters on the maria must have formed *after* the volcanic overflow. Similarly, if we see two craters that overlap, we can distinguish which crater formed first and which one formed second. The crater that formed later will lie on top of the older crater (Image A). We say that the younger crater has *modified* the older crater. We can say that the craters lying on top of the maria modified the maria. This method of relative dating is called *stratigraphy*. Stratigraphy only works if features overlap one another, but can we infer relative ages for features that do not overlap? In some cases, we can. *Morphology* is the study of shapes. With time, the process of erosion will slowly wear down the normally sharp features of lunar craters.

There is no atmosphere on the moon, nor is there any water, so the typical erosion processes that we experience on the earth are not present on the moon.

However, there are more gradual erosion processes on the moon. These erosion processes include slumping, thermal fatigue as lunar rocks heat and cool each lunar day,[5] radiation damage, and cosmic ray strikes. More dramatic are additional impacts. The impacts of larger bodies are relatively rare, but there is an appreciable flux of micrometeoroids on the moon. Each of these impacts, as well as the other erosion processes, will slowly round down the normally sharp relief of crater walls. Therefore, older craters appear more rounded and less sharp than younger craters. Of course, this assumes a relatively uniform erosion rate throughout the lunar surface, but this seems reasonable.

What about the rate at which new craters form? Most people assume that the rate was higher in the past than it is today. If the solar system started with a finite number of meteoroids and they aren't being replaced, then as meteoroids slam into the lunar surface, or any other surfaces in the solar system, the number of potential impactors decreases, so the cratering rate must decrease. How does one explain the very large meteoroids that formed the impact basins? Those impacts occurred relatively late, so where were they earlier on? Surprisingly, evolutionists do not have an answer to that question. Planetary scientists have worked out a cratering history for the moon with two intense episodes. The first, called the *early heavy bombardment* (EHB), resulted in the most of the craters in the lunar highlands. The second, called the *late heavy bombardment* (LHB), produced the impact basins. There has been relatively little cratering since the LHB and the subsequent volcanic overflow that filled the impact basins. Again, within the evolutionary paradigm it is a complete mystery where the very large meteoroids of the LHB came from.

Is there a creationary theory of the history of the moon? Notice that the history of the moon briefly outlined here assigned only relative ages, not absolute ages. Absolute ages may be determined only by additional information or assumptions. In the evolutionary worldview, the EHB lasted from the moon's formation 4.5 billion years ago until about 4.0 billion years ago. The LHB lasted from 4.1 to 3.8 billion years ago. Volcanic overflow of the impact basins commenced anywhere from 100 million to 500 million years after the formation of the impact basins. Biblical creationists can accept the relative ages, but would assign different absolute ages to the events of the moon's history.

---

[5] The day/night cycle on the moon is a month.

The model favored by the author is that much of the EHB corresponds to the formation of the moon on Day Four of the Creation Week. The biblical date of creation is shortly before 4000 BC,[6] so the EHB would have been more than 6000 years ago. Because the creation of astronomical bodies took place on a single day, all or nearly all of the EHB must have happened within one day. Obviously, this was a much more rapid process than evolutionists would believe. The LHB corresponds to an event related to the Flood. There are a number of astroblemes (fossil impact craters) found in sedimentary layers around the earth. If most of these sedimentary rocks were laid down by the Flood, then it follows that there was significant bombardment of the earth at the time of the Flood. The impacts must have played some role in God's judgment of the earth at that time, perhaps related to a triggering mechanism for the Flood. A swarm of large meteoroids must have entered the inner solar system and targeted the earth at that time.

Being so close to the earth, the moon would have been struck as well, as what the military calls collateral damage. The distribution of the lunar maria offers some clues about this event. The near side of the moon is about half-and-half maria and highlands. However, the lunar back side is about 95% highlands. Why is there such dimorphism between the lunar hemispheres? Interestingly, the maria of the moon's near side are clustered in the northern quadrant, leaving the southern quadrant that faces the earth mostly highlands. If the LHB impacts were randomly distributed over some considerable time, one would expect a much more random distribution than this. Such tight clustering of the impact basins and the subsequent volcanic overflow suggests that the direction that the large meteoroids followed was very directed. Furthermore, the fact that the lunar far side escaped nearly unscathed suggested that the bombardment was very brief, perhaps occurring over a period of less than a month. If the bombardment were spread over much time, the moon's rotation would have rendered even a unidirectional bombardment divided between both hemispheres. Some may object that there are some large impact basins on

---

[6] Exactly how much before 4000 BC the creation was is the subject of some debate. Some people today prefer Ussher's 4004 BC date, but others do not. It is unlikely that such precision in the biblical date of creation is possible. Furthermore, Ussher endorsed a short sojourn in Egypt, but a long sojourn would add nearly two centuries to the chronology. Finally, the LXX chronology would add more than a millennium to the chronology. At any rate, it is not possible to fix a biblical date of creation earlier than about 5500 BC.

the lunar surface that are not related to maria, and that some of these are on the lunar far side. However, these impact basins are not readily visible, because they have been heavily modified by subsequent cratering. This suggests that these impact basins are older than the LHB, probably placing them into the EHB. Why they did not produce volcanic overflow is not known.

Not all biblical creationists agree with this model. Some creationists object to having any craters on the moon prior to the Fall of man. To them, craters speak of judgment and fall far short of the perfect world that God created. This idea of perfection arises from the "very good" summary description of the creation found in Genesis 1:31. However, does "very good" equate with perfection? And what would perfection mean in this context? The "very good" of Genesis 1:31 has at least three connotations. First, it refers to the completeness of creation. The initial state of creation was described in Genesis 1:2 as being unformed and unfilled ("without form and void" in the KJV). We see over the six days of the Creation Week God's care in shaping and then filling the world, with the crowning achievement being the creation of man and his place of habitation (cf. Isaiah 45:18). God could have created everything in an instant, but instead He ordained a process whereby over six days of work He brought about the created universe (Exodus 20:11). Unfortunately, many biblical creationists fail to appreciate the role that processes played on each day of the Creation Week.

The second connotation is related to the first. Everything was "very good" in the sense of conforming to the purpose for which God created it. This is especially relevant because the purposes for the heavenly bodies are expressly mentioned in the Day Four account. They exist to give light on the earth, to mark the boundaries of day and night, and to function as indicators of special days and times.

The third connotation of the very good state of the creation is that sin had not yet entered the world. In this sense there was a moral goodness to the world. The introduction of sin in Genesis 3 changed everything. In addition to the directly enumerated effects of the Fall (Genesis 3:14–19), there were more general effects. Not only did death enter the world (cf. Romans 5:12–21; 1 Corinthians 15:20–49), but the taint of man's sin has affected all of creation (Romans 8:13–23) as well. This is why the very creation will one day be redeemed by means of its destruction by fire followed by the recreation of a new heaven and a new earth (Isaiah 65:17; 66:22; 2 Peter 3:10–13; Revelation

21:1). The Fall brought about the loss of this moral goodness, but the new creation will restore moral goodness to the world.

Notice that the word *perfection* is not found in the Scriptures related to this discussion. It is not that the word *perfection* does not convey the proper idea, but rather that perfection can have many connotations not intended by the biblical passages. The subtle substitution of other meanings commits the informal fallacy of equivocation. For instance, nearly all solids have a crystalline structure. A crystal is an orderly array of regularly spaced units, with a unit being an atom, ion, or molecule. Since atoms, ions, and molecules are very small, every crystal contains a huge number of constituent units. Any crystal will have at least a few sites where there are slight deviations from the orderly array. Physicists who specialize in the study of crystals call these sites *crystal defects*. In reality, perfect crystals, in the sense that they contain no defects, do not exist. Since the 1960s a common belief among biblical creationists has been that the original creation contained only perfect crystals, but that the introduction of sin induced a multitude of crystal defects, rendering all crystals less than perfect. While this sounds plausible to many biblical creationists, this argument commits the informal fallacy of equivocation. There is no moral reprobation associated with crystal defects, for they merely are a departure from the normal pattern associated with crystals. To insist that crystal defects somehow would have violated God's pronouncement of "very good" on creation amounts to assuming far more than the text requires.

Related to this teaching is insistence that the second law of thermodynamics came into existence at the time of the Fall. The second law of thermodynamics was formulated to describe the direction that heat (or energy) flows. The second law has been generalized to describe the amount of disorder in a system. Some biblical creationists have taken this generalized form of the second law and attempted to equate it with the physical effects of the Fall. This, too, has been attractive to many biblical creationists, but it raises more problems than it solves. Remember that the second law of thermodynamics originally described the direction that energy flows. Absent the second law of thermodynamics, the sun would not have shone. The same is true of the moon, stars, and other astronomical bodies, yet there is good evidence that Adam saw these. Digestion of food is likewise an excellent example of the second law of thermodynamics. We know that Adam and Eve ate food prior to the Fall (Genesis 1:29–30; 2:9,16–17), so presumably there was a need to digest food. When faced with these objections, those who support the notion that the second law of thermodynamics came into existence at the time of the Fall further compound the problem by piling conjecture upon conjecture to salvage their theory.[7]

Fortunately, this wild speculation greatly expanding the physical effects of the Fall beyond those required by Scripture has waned among biblical creationists in recent years. However, the related resistance to considering the possibility of craters as part of the creation remains. There were many processes that occurred during the Creation Week. God formed man from the dust of the ground (Genesis 2:7). God formed the land and flying creatures in a similar manner (Genesis 2:19). On Day Three, God caused plants rapidly to grow out of the ground as well (Genesis 1:11–12). Also on Day Three, God made the waters to gather in one place and made the dry land appear (Genesis 1:9). On Day Two, God made the rāqîaʿ ("expanse," or "firmament" in the King James Version) to separate the waters below from the waters above. All of these creative acts strongly imply processes. However, these processes were not random, nor were they slow. Rather, these processes were rapid and directed.

Given this pattern of rapid, directed processes that played out on other days of the Creation Week, might not have God used a similar mechanism on Day Four in making the heavenly bodies? If God assembled the astronomical bodies on Day Four, might there be any evidence of at least the final stages of the assembly on some bodies—such as on the moon's surface? Could at least some of the craters on the moon represent evidence of this assembly on Day Four? If not, then what kind of surface would one expect on the lunar surface? The moon has no need for vertical segregation as on the earth to separate the seas from dry land. If one keeps reasoning in this manner, soon one ends up with the moon originally being a perfectly smooth sphere. Ultimately, the same reasoning would have to apply to other astronomical bodies, such as the planets, to reach the same conclusion. Perfect spheres among the heavenly bodies may appeal to many people as conforming to

---

[7] When raising this objection, the author has heard the response, "We don't know what physics was like before the Fall," followed by hypothesizing how things might have worked then. Ironically, this tack is inconsistent. The insistence that the second law of thermodynamics was not present before the Fall is a statement that we do know what physics was like before the Fall.

THE LESSER LIGHT TO RULE THE NIGHT: THE MOON, OUR NEAREST NEIGHBOR   119

their expectation of an initial perfect creation. But this is not a new idea. The ancient pagan Greeks believed that the heavenly bodies—particularly the sun and moon—were perfect spheres. Their reasoning was that there was a dichotomy between the heavenly and terrestrial realm. While the earth obviously was imperfect, the heavenly realm was perfect. And what was a more perfect shape than the circle? A circle is a two-dimensional version of this perfection, while a sphere is a three-dimensional version. Ergo, heavenly bodies were spheres, and they exhibited uniform circular motion.[8] So, this idea is pagan, not biblical.

If one does not allow any craters prior to the Fall, then when did craters come about? One of the more common answers that biblical creationists have given to that question is that craters formed due to judgment associated with the Flood. However, in recent years some creation scientists have estimated how many meteoroid impacts this would have involved at the time of the Flood. Both the number of impacts required and the energy involved is staggering. If the earth shared in this intense bombardment, then the effects on earth would have dwarfed those of the Flood itself. As one creation scientist quipped, "We have the account of Noah's Ark, not Noah's bunker."[9] Consequently, support for relegating nearly all lunar impacts to the Flood has waned considerably in recent years.

Additionally, lunar ghost craters reveal something about the age of the moon and the cratering rate over time (Image B). Ghost craters are found in the lunar maria. They are the faint outlines of craters that have been covered or mostly covered by lava. Obviously, a ghost crater must have formed after the impact basin in which it is located was excavated, or else the formation of the impact basin would have obliterated the crater. But the crater must have formed before the volcanic overflow, or else the crater would lie on top of the lava flow instead of underneath. If one assumes that lunar cratering occurred over millions or billions of years, that fixes a certain cratering rate. However, if one assumes that craters accumulated in the relatively short time span of biblical creation, then the cratering rate must have been much greater. There is a strong one-to-one correspondence between impact basins and volcanic overflow on

---

[8] Incidentally, these constraints led to the Ptolemaic model.
[9] Steve Austin said this during a panel discussion on impacts at the Seventh International Conference on Creationism in 2013 in Pittsburgh, Pennsylvania.

the lunar surface, so the two must be related. It would seem that if the volcanic overflow used the fractures produced by the formation of the impact basins as conduits to reach the surface, then the overflow likely happened rapidly after the impact basins formed. However, most planetary scientists think that, depending upon which mare is under discussion, between 100 million and 500 million years transpired between those two events. Why do they think so much time elapsed between the two events? It is because their assumption of the rate of crater formation follows their assumption of billions of years for lunar history. That cratering rate would not allow for the two events to happen in rapid succession. However, in the much higher rate of crater formation implied by the creation model, this is no problem—one can have the volcanic overflow rapidly follow the formation of the impact basins and account for the density of ghost craters that we see. Therefore, lunar ghost craters conform better to the biblical creation model.

Where Did the Moon Come From?

According to Genesis 1, God made the moon on Day Four, three days after He made the earth. Of course, evolutionists do not agree with this. There is an evolutionary origin scenario for the solar system (the next chapter has a brief discussion of it), and the moon's origin is explained as part of this theory. By the 1960s there were three different theories about the moon's origin. These were:

- The earth and moon formed together (co-creation theory)
- The moon spun off from the early earth (fission theory)
- The moon formed elsewhere and was captured by earth's gravity (capture theory)

The 1960s witnessed the Apollo moon-landing program. At the time, everyone understood that the United States went to the moon for nationalistic reasons, but much of the justification centered on the science that a manned landing on the moon could accomplish. Part of this justification was to collect data to determine which of these three theories of lunar origin was correct. By the 1970s, the data collected on the moon, data assembled from other means, and physical considerations clearly showed that none of the three theories was correct.

For about a decade, there was no clear alternate theory, but by the 1980s a synthesis of elements of the three basic theories began to emerge as the favored model. In this model, the earth formed first, but shortly thereafter suffered a glancing collision by an object about the size of Mars. A key component of this theory is that this was a glancing rather than head-on collision. According to this theory, a glancing collision would have permitted much of the core of the impacting body to sink into the earth's core. Such a glancing collision, also would have caused a mixture of the mantles of both the early earth and the impacting body to eject into orbit around the earth. Presumably, the moon coalesced out of the debris in orbit around the earth, whereupon the moon's orbit gradually increased in size to place the moon where it is today (discussed later in this chapter). This collision supposedly took place 4.45 billion years ago, just 100 million years after the earth formed. There are differing variations on this basic theme.

As previously mentioned, most of the planets have satellites. For a long time, it was thought that a single explanation probably could account for the origin of most of these natural satellites of the planets. However, it now appears that several mechanisms might need to be invoked. For instance, many of the larger satellites in the solar system are thought to have formed by the co-creation theory, while many of the smaller ones likely were captured asteroids. But the evolutionary scenario for the moon's origin may be unique. This is because the moon has some unique properties. First, the moon's mass is large compared to the earth. As stated at the beginning of this chapter, the earth is 81 times more massive than the moon. For all other satellites in the solar system, the ratio is much more than a thousand to one. Given the much closer match between the masses of the earth and moon (and even closer match with regard to their diameters), the earth-moon system sometimes is referred to as the double planet.

But the moon's orbit is unique as well. Most of the larger satellites orbit in the equatorial planes of their planets (the two exceptions are Triton and Nereid, the two largest satellites of Neptune; I will discuss these in the next chapter). Many of the remaining smaller satellites also orbit close to the equatorial planes of their planets. The smaller satellites that have orbital planes highly inclined to the equatorial planes of their planets are thought to have been captured, though some of the ones that have orbits close to their planets' equatorial planes may have been captured too. However, the moon orbits close to the earth's *orbital plane around the sun*. No other satellites in the solar system do this (Table 1). This is powerful evidence that the moon's origin is unique. Few evolutionists seem to have grasped the significance of this fact. Evolutionists generally assume that there is nothing particularly special about the earth, including life on the earth. This is because in the evolutionary worldview, there is no design. If one were to admit that the earth has any special status, then that could imply that the earth is designed, which in turn could lead to the possibility of creation rather than evolution.

Table 1. Satellite inclination relative to planet's equatorial plane.

Planet	Tilt of Planet (°)	Satellite	Orbital Inclination (°)
Jupiter	3.12	Io	0.04
Jupiter	3.12	Europa	0.47
Jupiter	3.12	Ganymede	0.21
Jupiter	3.12	Callisto	0.51
Saturn	26.73	Dione	0.02
Saturn	26.73	Rhea	0.35
Saturn	26.73	Titan	0.33
Saturn	26.73	Iapetus	14.72
Uranus	97.86	Ariel	0.3
Uranus	97.86	Umbriel	0.36
Uranus	97.86	Titania	0.14
Uranus	97.86	Oberon	0.10
Neptune	29.58	Triton	157.34
Pluto	119.6	Charon	0.00

Is there evidence of design in the moon's unique orbit? Yes. In the previous chapter I discussed the earth's tilt as the cause of the seasons. The earth's tilt does not remained fixed at one value. There are forces at work in the solar system that gradually change the amount of the earth's tilt. Over even a few human lifetimes, the effect is modest, but it does add up. If these forces were left unchecked, over many thousands of years the earth's tilt would vary over the range of 0° (no tilt) to 90° (maximum tilt). Obviously, this would not be good for living things. During times of no tilt, there would be no seasonal variations in temperature and rainfall. This would be bad enough, but during times near maximum tilt, the consequences would be disastrous. Much of the earth's surface simultaneously would be in both the tropics and Arctic or Antarctic. Clearly, this would be problematic for all living things.

But there is no worry, for the earth's tilt does not change wildly in this manner. Why? Recall that the moon has a relatively high mass compared to the earth, and the moon orbits in the ecliptic plane, not in the earth's equatorial plane. These two factors combine to stabilize the earth's tilt, so that the tilt varies only over about a 2° range. If the moon did not exist, or if it had much less mass, or if the moon's orbit were any different, this stabilization would not occur. This is the only planet in the solar system on which this mechanism works, and it is the only planet where it matters, because it is the only planet where life exists.

Lunar Dust

When an impact occurs on the moon, debris is scattered for some distance around. This happens for all levels of impacts. There is a huge influx of micrometeoroids onto the moon. As the name suggests, micrometeoroids are very small, so the crater formed and the debris scattered by an individual impact is very small. However, micrometeoroids contribute much collectively. For instance, micrometeoroid impacts probably are one of the major erosion processes affecting lunar features. Micrometeoroids burn up in the earth's atmosphere, so they do not reach the earth's surface. Therefore, assessing the flux of micrometeoritic material is difficult. The earliest measurements in the late 1950s indicated a large influx. Since the lunar surface has no atmosphere to protect it, scientists expected the micrometeoritic dust that had accumulated on the moon's surface over billions of years would be very deep. Most estimates were within the range of hundreds of meters. Some scientists feared that the particles would be very tiny, which could cause lunar landers to sink far into the dust. Therefore, the Lunar Excursion Module (LEM) that the Apollo astronauts used to land on the moon had large inverted dishes on its feet. It was hoped that the large area would spread out the weight of the LEM and prevent it from sinking into the dust.

The Apollo astronauts found that the lunar surface was covered with a very thin layer (less than an inch) of fine dust. Since the dust on the lunar surface was far less than expected, many biblical creationists thought that it indicated that the moon must be far younger than billions of years. But is it? The earliest measurements of micrometeoroid flux were indirect. Later measurements were direct—such as by means of leaving a surface exposed to the environment of space while aboard an orbiting spacecraft for several years, whereupon the surface was retrieved and examined for tiny impacts. These direct measurements proved to be much lower than the early estimates of the micrometeoroid flux. With the new measurements, one would expect only a thin layer of lunar dust rather than a thick layer. Therefore, lunar dust is not a good argument for a creation that is only thousands of years old. Unfortunately, there are some biblical creationists who still attempt to use this argument.[10]

Transient Lunar Phenomenon

Transient lunar phenomenon (TLP) are short-lived changes in the appearance of the lunar surface. TLPs can be changes in brightness or color. Most observations of TLPs are telescopic, but there have been some naked-eye TLP observations. For instance, the most famous TLP report was in 1178, when five English monks saw a catastrophic event while looking at a thin waxing crescent moon. In 1787, the famous British astronomer William Herschel reported seeing through his telescope three red glowing spots on the dark part of the moon. In the modern era, the most famous TLP case is the half-hour glow in the region of the Alphonsus Crater that Russian astronomer Nikolai Kozyrev reported in 1958. Kozyrev managed to record two spectra of the event that contained emission lines due to carbon molecules.

TLPs are very controversial among astronomers and planetary scientists. There are many proposed explanations. For those who believe that TLPs are real events on the lunar surface, there are two major schools of thought. One proposal is that some TLPs are evidence of volcanic activity on the moon. Herschel and Kozyrev thought that the events they famously observed were volcanic eruptions. The other major suggestion is that TLPs are impacts. This has been a common explanation for the 1187 event witnessed by the English monks. Why are other scientists slow to embrace these explanations? Large impacts are relatively rare events, so the probability of anyone witnessing one is very low. Most scientists think that the moon is geologically dead; so with

---

[10] For more information, see see A. Snelling and D. Rush, "Moon Dust and the Age of the Solar System," *Creation Ex Nihilo Technical Journal* 7, no. 1 (1993): 2–42.

no significant internal heat, there can be no volcanic activity on the moon. The major reason to discount volcanic activity on the moon today is that if the moon is billions of years old, it ought to be cool inside. Any primordial heat would have dissipated long ago, and the moon lacks heavier elements that tend to be radioactive, which precludes radioactivity as a possible source of heat. Clearly, both explanations are of interest to biblical creationists. If it can be shown that the moon is volcanically active, then it would indicate that the moon is far younger than generally thought. Therefore, a volcanically active moon would be, by implication, an indicator of recent creation. Noticeable impacts on the moon today could be used to indicate a recent creation, because the rarity of large impacts today would not be expected in a solar system that is billions of years old.

If the moon is not volcanically active, and if noticeable impacts are very rare today, then how can one explain TLPs? Most explanations involve the earth's atmosphere. One suggestion about the 1187 event is that the Monks saw the shattering of a meteor nearly head-on in the earth's atmosphere that was in the same direction of the moon. This would explain why no one else seemed to have noticed the event. Some have suggested that documented strong aurora activity at the time of Herschel's discovery may have been what he saw. Turbulence in in the earth's atmosphere causes astronomical bodies to appear to flicker. This flickering is what causes stars to twinkle. When viewed with high power, images of the moon and other objects can roil because of the turbulence. Hence, bright spots on the moon can appear to vary in brightness and color when no real change has occurred. Interestingly, in recent years use of very sensitive high-speed cameras attached to large telescopes have captured images of what probably are impacts on the dark portion of the moon. However, these impacts are much more modest than those claimed as TLPs in the past. With these considerations, it is not clear exactly what TLPs are. Biblical creationists are advised to treat this subject with caution.

What Causes the Tides?

Anyone who has spent any time near the seashore is aware that the ocean levels fall and rise twice a day. Most people also know that this has something to do with the moon, but how does this work? Gravity compels the moon to orbit the earth each month. But by Newton's third law of motion (action-reaction), the earth orbits the moon each month too. Since the earth has 81 times more mass, the moon moves 81 times more than the earth does. The moon orbits more than 240,000 mi (644 km) from the earth, so the orbital distance of the earth is about 3000 mi (4828 km). The earth's radius is 4000 mi (6437 km), so each month the earth moves around a point that is 1000 mi (1609 km) below the earth's surface.

This orbital motion occurs as if the earth and moon were point masses, that is, point-like particles with all their masses concentrated in those points. However, the earth and moon are not point masses. Consequently, the earth pulls on the near side of the moon more than it pulls on the back side of the moon (Figure 4.3). In between these two extremes, the earth pulls on the moon by varying amounts. The moon does the same thing to the earth. We call this a differential force, because either body is pulling on different parts of the other by differing amounts. This differential force tends to stretch both the earth and the moon along the imaginary line connecting the centers of the two bodies. It is important to realize that this differential force produces a high tide on either side of the earth. Many treatments of the tides falsely state that it is the moon's gravity that pulls water upward, away from the earth. However, if this is the case, it is difficult to see why there is a tidal bulge on the opposite side of the earth. Furthermore, if the moon's gravity pulling water away from the earth is the reason for the tides, then why do we not see tides in small bodies of water, such as lakes, ponds, and even our bath tubs? The reason is that the force is parallel to the surface, and it is very feeble. Only when this feeble force is multiplied over the vast distances of the oceans does its effect pile up to noticeable values.

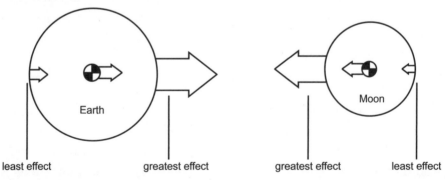

Figure 4.3. Illustration of gravitational effects of the earth on the moon and vice-versa.

Rocks can respond to such forces, so both the earth and the moon stretch slightly along the imaginary line connecting the two bodies. The amount of stretch is not much, just a matter of a few feet. The earth rotates rather quickly, so our rotation takes us through the two high spots on the earth, one on either side of the earth along the line connecting the earth and moon. In between, the earth's rotation takes us through two low spots. Consequently, each day the rocks beneath our feet heave up and down twice. However, it takes time for the rocks to react, and so they do not conform fully to the shape that the differential force is imposing on the earth. Therefore, anything free to move around on the earth's surface, such as very large bodies of water, make up the difference. This results in the twice daily rise and fall of ocean level. During one rotation of the earth, the moon moves nearly $1/30$ of its orbit around the earth. This carries the original high tide forward a bit, so it takes a little longer than 24 hours to go from one high tide to the corresponding high tide the next day. The average delay in the tides each day is 50 minutes, though the difference can be as little as 20 minutes and as much as 1½ hours.

The sun also raises tides on the earth. Even though the sun has far more mass than the moon does, it is much farther away. Therefore, the sun's tides are only about half as strong as the moon's tides are. Since the sun and moon usually do not line up, they produce high and low tides at different locations. When at new or full moon, the tidal effects do line up, and the solar and lunar tides reinforce one another. This produces the highest high tides and the lowest low tides, so the difference between high and low tide is at its greatest. We call this *spring tide*. This has nothing to do with the spring season, but rather the name stems from the large leap that the water level takes from low to high tide. At first and third quarter moon the lunar and solar tides oppose one another, with the sun producing low tides where the moon is producing high tides, and vice versa. Since the moon's tides are stronger, it wins out. But this is the lowest high tides and highest low tides, with the smallest difference between low and high tides. We call this *neap tide*. Each month the tides go through two cycles of spring and neap tides.

There are other effects that complicate this rather simple picture. One significant effect is the role that local geography plays. Local irregularities in coastlines can delay the timing of the tides, and it can diminish or accentuate tidal differences. The Bay of Fundy in Nova Scotia is famous for its unusually high tidal differences. The Bay of Fundy is shaped like a funnel that magnifies tidal differences. Another important effect is the earth's rapid rotation. It takes time for water to move sufficiently to produce the tides. From the simple description here, it would appear that high tide always would be when the moon is highest in the sky (or lowest below the horizon) and that low tide would be when the moon is on the horizon, either rising or setting. However, the earth's rapid rotation carries the tidal bulges forward so that tides are delayed a few hours with respect to the moon's location. That is, high tide occurs a few hours after the moon is highest (or lowest) above the horizon, and low tide is a few hours after moonrise or moonset.

This last factor has a very subtle short-term effect that is profound over great time. The earth's rapid rotation carries the earth's tidal bulge ahead of the imaginary line between the earth and moon. This bulge provides a sort of handle that the moon's gravity can pull on. Either end of the bulge is accelerated toward the moon. The bulge that is farther from the moon is pulled forward, which acts as a torque to speed the earth's rotation. However, the bulge that is closer to the moon is pulled backward, acting as a negative torque that slows the earth's spin. Since the closer bulge has a stronger force, the braking torque dominates over the accelerating torque, with the net effect being that the earth's rotation is slowed. We can measure the rate of tidal braking from historical records of where total solar eclipses were visible. If we compute where these eclipses ought to have been visible based upon the assumption of no tidal braking, we find that the actual locations from observations are systematically shifted eastward in longitude. From this shift, we determine that the day is lengthening at a rate of 0.0016 seconds/century.

But that is not the end of the story. Newton's law dictates that for every action there is an equal and opposite reaction. The tidal bulge on the earth pulls back on the moon, accelerating the moon forward in its orbit. This acceleration has the effect of lifting the moon to a higher orbit, so the moon is slowly spiraling away from the earth. The Apollo astronauts left special mirrors on the lunar surface that allow us to measure the rate at which the moon's orbit is increasing in size. From time to time, astronomers use large telescopes to send powerful laser beams to the moon and to collect light reflected off the mirrors. The time delay between when the pulse is sent and the detection of the reflected pulse precisely measures the distance to the moon. At this level of precision, the moon's orbit is very complicated, with all sorts of subtle tweaks occurring. When these effects are removed, the remainder is the change in the moon's distance. Astronomers consistently find that the moon's orbital radius is increasing at a rate of about four centimeters per year. Over a thousand years, the moon moves an average of 40 m (130 ft) farther away. Even over a million years, the change in distance would amount to only 40 km (25 mi), as compared to a total distance of nearly 400,000 km (250,000 mi).

While this increase in the moon's orbital radius is reasonably linear over the short run, over long spans of time it is not linear. From the underlying physics, we know that the rate of lunar recession is inversely proportional to the sixth power of the distance. This is a very steep function with respect to distance, which means that the rate of lunar recession is highly dependent upon the moon's distance. In the past, when the moon would have been closer to the earth, the rate of lunar recession must have been far greater than the current rate of 4 cm (1.5 in) per year. Using relatively simple calculus, one can show that a little more than a billion years ago, the moon would have been nearly in contact with the earth. When so close to the earth, the earth's tidal force would have shredded the moon. About a billion years ago, the tides raised on the earth's oceans by the moon would have been a mile high. No one believes that this ever happened, because no one believes that the moon was this close to the earth then. It is clear from this that the moon could not have been orbiting the earth for a billion years, let alone nearly 4.5 billion years. However, if the moon is only thousands of years old, as biblical creationists believe, then this is not a problem. Therefore, the tidal evolution of the earth-moon system is an evidence that the earth-moon system is not nearly as old as generally thought.

How do those who believe that the world is billions of years old respond to this? They argue that even apart from its strong dependence upon distance, tidal evolution is not constant. This is true, because tidal braking depends on the arrangement of continental masses as well. If the continents had a different distribution, then the rate of tidal evolution would be different. This is difficult to model, even if we knew the distribution of the continents over time, which we do not. Therefore, the response amounts to asserting that we happen to live at a time of unusually high tidal interaction between the earth and moon. Even further, this includes a supposition that this unusually high rate of tidal evolution has prevailed for half a billion years or more. Presumably, the earth and moon were in a sort of tidal lock for much of the preceding four billion years. Is there any evidence for this tidal locking? No, but there have been computer simulations that ostensibly show how this might have happened. Increasingly, this is the way in which much of science happens: computer simulations are substituted for data.

As the moon raises tides on the earth, the earth raises tides on the moon. Being much more massive, the tides on the moon are much stronger than those on earth. This raises a bulge in the rocks of the moon, and as with the moon acting on the earth's bulges, the earth produces a torque on the moon, slowing its rotation. As previously discussed, most astronomers think that this is why the moon has synchronous rotation. Tidal braking has slowed the moon's rotation so that it rotates and revolves at the same rate. Synchronous rotation is common among the natural satellites of the other planets, and presumably is caused by tidal braking as well. This mechanism takes time, possibly more time than the biblical creation model would allow. If this is the case, then creationists need to find another reason (or purpose) why the moon and most natural satellites have synchronous rotation.

Eclipses

There are two types of eclipses, lunar and solar. A solar eclipse occurs when the moon passes between the earth and sun, so that the moon blocks out the sun, preventing at least some of the sun's light from reaching the earth. Alternately, one could say that the moon's shadow passes over the earth. Obviously, a solar eclipse can happen only during the moon's new phase. A new phase happens once per month, but solar eclipses do not happen this often. How can this be? The moon's orbit is tilted a little more than 5° to the ecliptic. Unless the moon is near a node at the time of new moon, the moon's shadow will pass above or below the earth, entirely missing the earth.[11]

The moon's shadow has two parts, the *umbra* and *penumbra*. Please refer to the accompanying diagram, but note that the diagram is not to scale with regard to the sizes of the bodies or their separations (Figure 4.4). A line drawn from the top of the sun past the top of the moon, and a line drawn from the bottom of the sun past the bottom of the moon will intersect at a point opposite

---

[11] The nodes are the two points where the moon's orbit crosses the ecliptic, the plane of the earth's orbit.

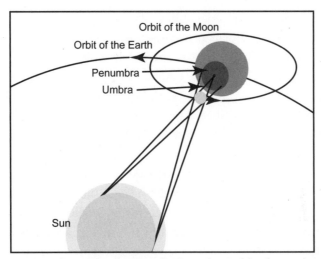

Figure 4.4. Lunar umbra and penumbra (Wikimedia. Public domain).

the sun from the moon. In the diagram these two lines, along with the diameter of the moon, form a triangle. From within the triangle, no portion of the sun would be visible, so this region would be very dark. This is the *umbra*, coming from the Latin word for *shadow*. This is a two-dimensional representation, while the actual situation is three-dimensional. While the umbra appears as a triangle in the diagram, its three-dimensional shape is a cone, with the moon's circumference forming the base, and the apex at the location where the two original lines intersect. If we draw lines from the top of the sun past the bottom of the moon and from the bottom of the sun past the top of the moon, those lines will intersect before reaching the moon. From that point of intersection, the two lines will diverge beyond the umbra. These two lines represent a cone extending indefinitely beyond the apex of the umbra. This cone is truncated by the moon's circumference that is the base of the umbra. This truncated cone is the moon's penumbra, meaning *partial shadow*. From within the penumbra, the sun's light is only partially blocked out. From any location that is in neither the umbra nor the penumbra, the sun's light is fully present.

At the moon's average distance from the earth, the moon's umbra barely reaches the earth's surface. Therefore, the width of the umbra is very narrow, not more than a couple of hundred miles across, and usually less than that. For locations within the umbra, the sun completely disappears. This is a total solar eclipse. During a total solar eclipse, the sky gets dark, similar to dusk, and the brighter stars and planets become visible. The corona, the outermost layer of the sun's atmosphere, appears as a pearly-white halo extending a solar diameter beyond the eclipsed sun. Often, streamers in the corona trace out the sun's magnetic field. Around the sun's limb (edge), prominences—blood-red loops—are visible. If a white background, such as a sheet, is placed on the ground, one often sees fuzzy shadow bands move across for a few minutes near totality. No photographs do justice to the remarkable appearance of a total solar eclipse (Image C). Many people describe it as a very moving, even spiritual experience. Totality lasts at most seven minutes, but usually it is far less than that. It is important to note that during totality, there is no need of special filters to protect one's eyes from the sun's rays. However, this is true *only* during the brief totality.

If one is in the moon's penumbra, then the sun's light is only partially blocked out. This is a partial solar eclipse. A partial solar eclipse is *not* safe to look at without proper filters. While interesting, partial solar eclipses pale in comparison with total solar eclipses. The corona and prominences are not visible. It does not get dark. Unless the partial eclipse is close to total, no shadow bands are seen. The only effect is that a portion of the sun's disk is blocked, though it requires a special filter or projection of an image to watch even this. It appears as if a bite has been taken out of the sun. The moon's umbra typically makes a narrow swath across the globe, with a much wider swath on either side where a partial eclipse is visible. In addition, even for locations where the umbra passes, the total solar eclipse is preceded and followed by partial phases that last for more than an hour.

When the moon is near apogee, the point on its orbit farthest from earth, the moon's umbra fails to reach the earth's surface. For positions located beyond the apex of the umbra, most of the sun is blocked, except for a ring around the edge of the sun. This is an *annular solar eclipse*. Notice that the word here is *annular*, meaning a *ring*, because a bright ring of the sun is left visible. As with a partial solar eclipse, an annular eclipse cannot be viewed safely without proper filters or projection of the image. If an annular eclipse is very close to total, then a few things visibly experienced during a total solar eclipse are possible. For instance, shadow bands may be visible, and brighter planets, such as Venus, may become visible.

The sun is 400 times larger than the moon, but the sun also is 400 times farther away than the moon is. These circumstances mean that the sun and moon have the same angular diameter in the sky, about ½°. Thus, during a total solar eclipse, the moon just barely covers the sun. This has several interesting effects. First, it makes total solar eclipses possible. If the moon were much smaller or farther away from the earth, there would be no total solar eclipses. Second, total solar eclipses are spectacular, but if the moon were much larger or closer to the earth, they would not be. Third, total solar eclipses are very rare. A total solar eclipse takes place somewhere in the world every year or two, but they are visible in only a relatively small part of the earth. On average, at any given location on earth, a total solar eclipse happens about once every four centuries. There are more than 150 other natural satellites in the solar system, but none of them combine these qualities to produce such spectacular and rare solar eclipses. Either they produce no total solar eclipses, or the eclipses they produce are so frequent and grossly over total as to not be spectacular. The earth is the only planet where the wonder of total solar eclipses can happen, and it also is the only planet where it matters, for it has inhabitants who can appreciate this fact. Therefore, the circumstances that combine to produce total solar eclipses is an indication of design.

The earth also has an umbra and penumbra. However, being a larger body, the earth's umbra is larger than the moon's umbra. At the moon's distance from the earth, the earth's umbra exceeds the diameter of the moon. The next figure shows what the earth's umbra and penumbra look like near the moon (Figure 4.5). The intersection of the cones of the moon's umbral and penumbral with a plane perpendicular to the axes of the cones is two concentric circles. The inner circle is the umbra, and the outer circle is the penumbra. A total lunar eclipse happens when the moon is entirely immersed in the umbra. One might expect that during a total lunar eclipse no sunlight reaches the moon, so that the moon would be completely dark. However, the earth's atmosphere bends light around the edge of the earth so that the umbra is not completely dark. This is the same effect that causes the sky to be light for some time after sunset (and before sunrise). And, as at near sunset and sunrise, red light is passed more into the earth's umbra than other colors. Therefore, the totally eclipsed moon often has a red or orange hue. However, the color is quite varied from eclipse to eclipse. Sometimes the moon appears almost black, while other times it can be

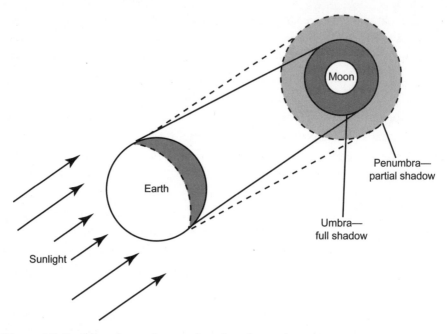

Figure 4.5. Earth's umbra and penumbra viewed near the moon.

golden. The maximum duration for a total lunar eclipse is about an hour and 40 minutes, though most total lunar eclipses are shorter than this.

A partial lunar eclipse occurs when the moon enters the earth's umbra but fails to be completely immersed. The eclipsed part of the moon appears as a bite taken out of the moon. The light of the uneclipsed part of the moon is many thousands of times brighter than the eclipsed part. This makes it difficult to determine what color the umbra is, except when the partial eclipse is close to being total. A penumbral eclipse occurs if the moon misses the earth's umbra entirely but enters the earth's penumbra. A penumbral eclipse is easy to miss. All that happens is the moon is made slightly dimmer than normal, because it is receives less light from the sun while in the penumbra. However, the human eye has the amazing ability automatically to adjust light levels, so the moon appears as normal. If a penumbral eclipse is deep enough so that the moon's limb approaches the earth's umbra, it is possible to notice some dimming on the part of the moon closest to the umbra, but it is very subtle.

Whereas a solar eclipse is visible from only a portion of the earth's surface facing the sun, a lunar eclipse is visible from the entire nighttime half of the earth. Therefore, more people can see any given lunar eclipse. The result is that solar eclipses of all types are more common, but lunar eclipses are more often seen. As with solar eclipses, lunar eclipses happen only during a particular phase of the moon. Since the moon must be opposite the sun for the earth's shadow to fall on it, that phase must be full. And, as with solar eclipses, there is not a lunar eclipse each month when the moon's phase is suitable. Again, the inclination of the moon's orbit around the earth usually causes the moon's shadow to pass too high or too low for an eclipse to happen. Only if the moon is near a node during a full moon can a lunar eclipse occur.

Have We Landed on the Moon?

In 1961, President John F. Kennedy committed the United States to landing a man on the moon and returning him safely to earth. This bold initiative was announced only four years after the space age had begun, with the Soviet Union's launch of Sputnik, the first artificial satellite. This was the height of the Cold War, when it appeared that the United States was far behind the Soviet Union, and this was viewed as the best program to establish the West as a leader in technology and exploration. Despite the cost, most Americans overwhelmingly supported this effort. The dream was realized on July 20, 1969, when Neil Armstrong became the first human being to step out onto the surface of another world. Interest in going to the moon soon waned, and after six successful missions to land on the moon, the Apollo program (the name of the program that took astronauts to the moon) was canceled. The space program, both manned and unmanned, continues to today, though since the cancellation of Apollo it has never enjoyed the broad support that it did in the 1960s.

Even at the time of Apollo 11, there were a few people who doubted that we had landed on the moon, but they were an extremely small and silent minority. It is amazing that about 30 years after the Apollo program ended, that began to change. In the first decade of the twentieth century, several sources publicly began to doubt that we had landed on the moon. In 2001, the Fox Television Network broadcast an hour-long documentary, *Conspiracy Theory: Did We Land on the Moon?* The rise of the internet and social media have fueled belief that the Apollo moon landings were hoaxes. There are many thousands of videos on the internet promoting this view. Other venues of popular culture, such as television and movies, have contributed.

Why has the idea that we never landed on the moon become so popular? There is a tremendous amount of distrust of government and other public institutions today. Well into the 1960s, people were far more trusting. Discontent with the Vietnam War in the late 1960s caused people to question the government. A few years later, the Watergate scandal unfolded as the United States was extricating itself from that war. Both Watergate and the war changed the nature of journalism into the more adversarial profession that it is today. At the same time, there was much public fascination with alternate theories of reality. Much of this can be traced to suspicion that many people had of the Warren Report, the 1964 publication of an official investigation into the assassination of President Kennedy that had happened the year before. Many people doubted the conclusions of the Warren Report, which then left people to raise the question of who killed Kennedy. This naturally led to all sorts of wild speculations of conspiracies, which in turn resulted in the publication of many books. Oliver Stone's 1991 movie, *JFK*, influenced a whole new generation of young people into believing that there was some great conspiracy involved in the Kennedy assassination.

Also in the 1960s, people began to question whether the military was withholding important information about UFO sightings. The military conducted Project Blue Book (1952–1970), but there were many critics of the official conclusions. The UFO craze had already begun in 1947. One of the events in the summer of 1947 was the crash of a balloon near Roswell, New Mexico. The balloon had been part of a very secret program, so the military acted to suppress information at the time, but the nature of this program was declassified in the mid-1990s.[12] Nevertheless, the Roswell incident remained relatively obscure for three decades. However, the publication of several books beginning with *The Roswell Incident* in 1980 generated a huge amount of interest. Various fantastic stories about what happened in Roswell arose. Many of these stories involved the crash of a flying saucer and the deaths of its alien occupants. The military supposedly conducted autopsies and has kept

---

[12] The program was called Project Moghul and consisted of very sensitive microphones attached to high-altitude balloons to listen to detonation of atomic bombs in the USSR to monitor development of Soviet nuclear weapons. Being the only way to detect these tests, the US military understandably wanted to keep the true purpose of the balloon that crashed secret at all costs. The Roswell event occurred early in the Cold War, and the declassification happened after the Cold War ended.

the remains frozen ever since. This interest was what eventually prompted the military to declassify the events of 1947, but the critics quickly rejected the official explanation. The fascination with UFO sightings explains the popularity of the 1990s hit TV show, *The X-Files*. The premise of this show was that the government was covering up all sorts of odd things, such as flying saucers.

The 1999 movie *The Matrix* and its sequels have helped fuel belief in fantastic conspiracies. *The Matrix* imaged a world that is entirely virtual reality. Nothing is real, and so everything that we think is real is not real at all. Hopefully everyone realizes that this movie and many other movies and TV shows are fiction, but it appears that far too many people find such wild tales plausible. At one time, NASA was one of our most trusted institutions. But now millions of people seem to be convinced that NASA faked the Apollo moon landings. Once one accepts that idea, it is not hard to entertain the possibility that *everything* that NASA does and has done is fake. As discussed in the Appendix A, belief that NASA is a completely fraudulent agency is necessary to belief in the "snow globe earth," that is, the idea that the earth is flat and surrounded by a hemispherical dome. However, certainly not all Apollo moon landing deniers are flat-earthers.

Why did NASA supposedly fail to land on the moon? The deniers often do not clearly answer that question, so there are several possibilities. One possibility is that space travel is impossible. Those who believe in the flat earth would belong in this camp. One reason for not going to the moon that some lunar landing deniers claim is that it was too dangerous to send people to the moon. This claim is based upon early concern about the Van Allen radiation belts around the earth. The earth's magnetic field traps charged particles from the sun in a set of broad bands around the earth. In low earth orbit, danger that these particles represent is minimal, but a trip to the moon would present some risk. But would that risk necessarily be lethal? No. The astronauts passed through the most dangerous parts of the Van Allen belts very quickly, because they were traveling so fast (nearly 25,000 mph [40,000 kph]). Furthermore, the thin metal skin of the Apollo spacecraft offered some protection. Still, the Apollo astronauts did experience a level of exposure that increased their chance of harm, such as from cancer. Most astronauts considered this an acceptable risk. Another suggested possibility is that NASA fell behind schedule and saw that it could not make its commitment to reach the moon on the timetable laid out by President Kennedy, so they faked the moon landings to fulfill this goal to beat the Soviets to the moon. The problem with this scenario is that the Soviets certainly were convinced that the United States successfully made it to the moon and back. They could intercept all radio transmissions and telemetry that NASA was receiving, and so scientists and engineers in the Soviet space program could have determined whether NASA faked the landings.

What evidence do those who allege that the Apollo moon landings were faked offer? They offer analysis of many Apollo photographs, videos, and audio recordings that supposedly demonstrate that the lunar landings were faked. For instance, they point out that photographs taken by astronauts on the lunar surface show that illumination appears to come from two different directions rather than one direction. If the sun is the sole source of illumination, then there would only be one source of light, but if the landings were faked on a stage, then there could be more than one light source. However, there is no requirement that a faked landing necessarily would use two light sources. In fact, it is insulting to suggest that the clever people who supposedly faked all of this overlooked such a fundamental detail. Furthermore, the lines of shadows supposedly showing two directions of illumination have been drawn improperly. When the lines are properly drawn, they clearly show there is one source of illumination.

Another argument alleges that on videos, the United States flags that astronauts set up at their landing sites are seen to be waving in the breeze, but this could not happen on the moon, where there is no air and hence there is no breeze. One needs to realize that because there is no air on the moon, the flags normally would have limply hung on their poles. To make the flags visible and hence stand out, NASA framed the flags with a metal wire. The supposed waving in the breeze happened only when the astronauts twisted the poles back and forth to drive the poles into the thin lunar soil. Therefore, the waving merely was due to vibrations induced by the astronauts; when the astronauts stopped moving the poles, the flags stopped rippling. Most of the supposed proofs of the faked lunar landings easily can be dismissed in this manner.

Why are so many people taken in by such poor arguments? And why are so many Christians taken in by the supposed Apollo moon landing hoax? This would make a good sociological study. As previously discussed, the widespread distrust of public institutions today certainly must play a role. However, I fear that what I call "Christian Gnosticism" is at work here. Gnosticism takes many forms, but one of its essential elements is the belief that secret knowledge leads to enlightenment. By learning this secret knowledge, one allegedly can assume a higher spiritual plane. Many professing Christians today seem to thirst for some sort of secret knowledge, but this amounts to a shortcut to spirituality that dismisses the sufficiency of Scripture and attempts to bypass the pursuit of sanctification. Their attitude seems to be that they have found this knowledge, and if others listen to them, they too can achieve this secret knowledge. The type of secret knowledge promised takes many forms. To many Christians, such secret knowledge largely concerns end-time prophecy. I can count at least nine so-called "end of the world" prophecies that were supposed to have happened during my adult lifetime. At the time of this writing, the latest end of the world was supposed to coincide with the four total lunar eclipses that occurred at the time of Passover and the Feast of Tabernacles during 2014 and 2015.[13] As each predicted return of the Lord fails to materialize, instead of being disillusioned, many followers blithely move on to the next prediction. Other examples of Christian Gnosticism include the Bible code, numerology, and the gospel in the stars.[14] Christian Gnosticism also appears to be at work in modern geocentrism, in belief in a flat earth (see Appendix A), and in the denial of men having landed on the moon.

At the heart of human attraction to supposed secret knowledge is pride. People derive tremendous self-worth by having the good sense to have figured things out that most people fail to grasp or refuse to understand. It does not seem to matter to many people who are immersed in these ideas that what they think is true is in fact false. Once people have sunk deep enough into bizarre conspiracy theories, it is nearly impossible for others to reason with them using facts. To agree that they have been duped would be an explicit admission that they were not quite as smart and enlightened as they thought that they were. Thus, pride also keeps people trapped in their worlds of make believe.

---

[13] For more about the supposed "blood moons," see Danny R. Faulkner, *The Created Cosmos: What the Bible Reveals about Astronomy* (Green Forest, Arkansas: Master Books, 2016), 157–180.

[14] The so-called gospel in the stars is thoroughly debunked in Danny R. Faulkner, *The Created Cosmos: What the Bible Reveals about Astronomy* (Green Forest, Arkansas: Master Books, 2016), 257–318.

Deuteronomy 19:15 states that two or three witnesses are required to establish a matter. Two of the 12 men who walked on the moon later became Christians. Both men, Jim Irwin and Charlie Duke, wrote their stories. Read what Charlie Duke wrote:

> I used to say I could live ten thousand years and never have an experience as thrilling as walking on the moon. But the excitement and satisfaction of that walk doesn't begin to compare with my walk with Jesus, a walk that lasts forever.
>
> I thought Apollo 16 would be my crowning glory, but the crown that Jesus gives will not tarnish or fade away. His crown will last throughout all eternity (see 1 Corinthians 9:25).[15]

Yes, we did land on the moon. Christians who deny this are in the position of accusing two Christian brothers of lying about one of the biggest things that ever happened to them.

---

[15] Charles Duke and Dotty Duke, *Moonwalker: The True Story of an Astronaut Who Found that the Moon Wasn't High Enough to Satisfy His Desire for Success* (Nashville, Tennessee: Oliver Nelson, 1990), 280.

CHAPTER 5

Heavenly Wanderers:
Planets and Satellites of the Solar System

Introduction: What's in the Solar System?

Before wading into a discussion about the solar system, it would be best to define what the solar system is, along with a few other related terms. The solar system consists of the sun and the objects that orbit the sun under the influence of its gravity. The sun contains 99% of the mass of the solar system, which leaves only 1% of the mass for everything else. The "everything else" is dominated by the planets. What is a planet? For our purposes now, we can define a planet as a large body that orbits the sun. There are now nearly a million known objects that orbit the sun, but only eight of those are recognized as planets, so how large must an object be to be a planet? That question does not have a simple answer, so for now let me arbitrarily define Mercury to be the smallest mass that an object orbiting the sun must have to be considered a planet.

All but two of the planets have natural objects orbiting them (Mercury and Venus are the two exceptions). Astronomers call these objects satellites, though satellites often are commonly referred to as *moons*. As noted in the last chapter, the informal usage goes back four centuries to the invention of the telescope when astronomers first discovered that some of the other planets had satellites. This reflects the similarity that the satellites of other planets have with the earth's moon in that they orbit those planets in similar fashion to how the moon orbits the earth. Up to that point, the moon was viewed as a special object, and the discovery of satellites orbiting other planets appeared to undermine the moon's special status. Since the moon is the proper name of the earth's natural satellite, astronomers prefer to avoid confusion by not using the word *moon* to refer to the satellites of the planets. Two satellites, Jupiter's Ganymede and Saturn's Titan, are larger than Mercury,[1] so why are they not planets? By definition, a planet must orbit the sun directly. Therefore, since Ganymede and Titan orbit planets rather than the sun, they themselves cannot be planets and must be satellites.

Astronomers classify most of the remaining large number of objects that orbit the sun as *small solar system bodies* (SSSBs). The International Astronomical Union (IAU)[2] created this term in 2006 in recognition of the blurring distinction between the two groups of small bodies that were combined into this new classification: comets and asteroids. Heretofore, comets had been small, icy objects that typically had very long, elongated orbits. Comets typically spend most of their time in orbit near aphelion, far from the sun. Being such small objects, comets reflect very little sunlight, so comets normally are too faint to detect. However, a comet must reach perihelion once each orbit. Comet perihelia typically are very close to the sun, and during the brief time around perihelion passage, heat from the sun sublimes much of the ice, and the sun's radiation causes the released gases to glow, resulting in a tremendous increase in brightness. As a comet recedes from the sun, it rapidly fades into obscurity once again.

For a long time, the defining characteristic of comets appeared to be outgassing of material. This is very different from asteroids, which had been viewed as small rocky bodies, lacking ices and hence incapable of outgassing. Additionally, the orbits of comets and asteroids were recognized as being fundamentally different. As mentioned above, comets generally have highly elliptical orbits. The orbital planes of the comets can have any inclination to the ecliptic. These characteristics are very different from those of asteroids. Asteroids have orbits that are far less elliptical, making them nearly circular. Furthermore, their orbital planes are much less inclined to the ecliptic, generally less than 20°. These orbital characteristics are like those of the planets, which

---

[1] However, they have less than half the mass of Mercury. Obviously, the density of Mercury is greater than either Ganymede or Titan.

[2] Founded in 1919, the IAU is an organization of astronomers from all over the world. The IAU was formed for several reasons, one of the primary reasons being the establishment of nomenclature in astronomy. Hence, astronomers worldwide accept the designations made by the IAU.

is why astronomers prefer to call these objects *minor planets*, a practice that I will follow in the remainder of this book.[3] While minor planets share orbital characteristics with planets, they generally are too small to possess enough gravity to have pulled themselves into spherical shape as planets have.

By the twenty-first century, astronomers began to realize that the distinction between comets and minor planets was not so clear. With the increase in the number of large telescopes and improvements in instrument sensitivity, astronomers learned that many minor planets were exhibiting outgassing behavior more indicative of comets. This was not a problem for minor planets that had orbital characteristics more typical of comets, but eventually some minor planets with classic orbits of minor planets were observed to outgas. Furthermore, the composition of minor planets with orbits far from the sun made it clear that if they were to pass close to the sun, they too would outgas and probably would appear as comets. Therefore, it appears that the once clear distinction between comets and minor planets is not so clear anymore. Rather, the older distinction was based upon viewing the extremes of a spectrum of characteristics.

While the IAU was reclassifying comets and minor planets as SSSBs in 2006, it also dealt with the changing understanding of Pluto. Since its discovery in 1930, Pluto was viewed as a planet, though it was problematic even then. The main difficulty was, given its faintness, Pluto almost certainly was far smaller than any other planet. It was not until 1979, with the discovery of Pluto's first known satellite, Charon, that astronomers could determine the mass of Pluto (less than 5% the mass of Mercury). Mutual eclipses of Pluto a few years later permitted the first direct measurement of Pluto's size (less than half Mercury's diameter). Furthermore, in 2003 astronomers discovered Eris, a minor planet comparable in size to Pluto, but orbiting the sun even farther out. Originally, astronomers thought that Eris was slightly larger than Pluto, but now they think that Pluto is slightly larger. However, Eris appears to be more massive than Pluto. If Pluto is a planet, then why not Eris? And if Eris and Pluto are planets, why not other minor planets discovered in recent years that are nearly as large? There appeared to be a natural break between Pluto and Mercury (in terms of size and mass), but not between Pluto and the minor planets smaller than it.

---

[3] The one exception is the term asteroid belt. Use of this term dates to the mid-nineteenth century. There has been no effort to rename it the minor planet belt.

Consequently, Pluto was reclassified as a minor planet, and it received a number reflecting the order of its recognition as a minor planet. Therefore, Pluto is officially known as 134340 Pluto. While it was in the midst of making other changes, the IAU created a new subclass of minor planets for Pluto and other large minor planets: *dwarf planets*. A dwarf planet is a body that is too small to be a planet, yet has enough mass so that its gravity has pulled itself into a spherical shape as planets have. In addition to Pluto, at the time of this writing there are four other dwarf planets: 1 Ceres, 136108 Haumea, 136199 Eris, and 136472 Makemake. The number of dwarf planets is sure to grow with the inclusion of some already recognized minor planets and possibly the discovery of additional ones.

The reclassification of Pluto initially met with some level of popular opposition, though this has faded considerably. Apparently, many people were upset about losing a planet, but they did not realize that the first minor planet discovered (1 Ceres) in 1801 was hailed as a new planet at the time and was viewed as such for about a half century. Therefore, there is historical precedent in reclassifying an object from being a planet, though no one remains alive from the time a planet previously was demoted. Many biblical creationists were suspicious of the change in Pluto's status, fearing that there might have been some evolutionary agenda underlying it, but that is not the case.

Two Types of Planets

The accompanying table shows selected orbital and physical data of the planets, along with Pluto (Table 1). The first column of the table gives the names of the planets. The second column gives the orbital distance from the sun, expressed in astronomical units (AU). An AU is defined to be the average distance between the earth and sun. An AU is approximately 93,000,000 mi, or 150,000,000 km. Those numbers are very large, making comparison of orbital distances expressed in miles or kilometers difficult to grasp. However, the AU is a convenient way to see differences in orbital sizes. For instance, notice that the first four planets are closely spaced, but that the outer four planets are not. Furthermore, there is a natural gap that appears between the two groups of planets. We shall see that the inner four planets share characteristics, while the outer four planets share characteristics, but the characteristics of the two groups of planets are different. We call the inner four planets the *terrestrial*

Table 1. Orbital and physical data of the planets (and Pluto).

Planet	Distance from sun (AU)	Mass	Diameter	Density	Rotation Period (Days)	Number of Satellites	Rings
Mercury	0.387	0.055	0.383	5.43	58.6	0	No
Venus	0.723	0.815	0.949	5.24	-243	0	No
Earth	1.000	1.000	1.000	5.515	0.997	1	No
Mars	1.52	0.107	0.533	3.94	1.03	2	No
Jupiter	5.20	318	11.2	1.33	0.414	67	Yes
Saturn	9.58	95.2	9.45	0.70	0.444	62	Yes
Uranus	19.3	14.5	4.01	1.30	-0.718	27	Yes
Neptune	30.2	17.2	3.88	1.76	0.671	14	Yes
Pluto	39.5	.0022	0.186	1.86	6.39	5	No
	Jovian	Terrestrial	Terrestrial	Jovian	Terrestrial	Jovian	Terrestrial

planets, because they resemble the earth. The outer four planets are the *Jovian planets,* because they are like Jupiter. Distance from the sun is one major distinguishing characteristic: the terrestrial planets are closer to the sun, while the Jovian planets are much farther away.

The third column lists the diameter of each of the planets. However, as with orbital size, use of miles or kilometers would be difficult to grasp, so the sizes of the planets are expressed in terms of the size of the earth. As you can see, the other terrestrial planets are smaller than the earth, meaning that the earth is the largest terrestrial planet. On the other hand, the Jovian planets are much larger than the terrestrial planets. The fourth column lists the mass of each planet. Again, I have chosen to express this in terms of the earth's mass, because stating the masses in terms of kilograms would result in very large numbers, but stating the masses in the form of this ratio makes for easy comparison. Notice that the earth is the most massive of the terrestrial planets, but that the Jovian planets are far more massive than the terrestrial planets. The masses and sizes of planets can be combined into density, the amount of mass per unit volume. The fifth column lists the density of each planet, expressed in grams per cubic centimeter. Notice that the terrestrial planets have a higher density than the Jovian planets. This indicates that the two types of planets are made of different material. The densities of the terrestrial planets tell us that they are made of rocky material. However, the Jovian planets cannot consist of much rocky material, or their densities would be far higher. Instead, the Jovian planets probably are made mostly of hydrogen and helium. These elements typically are gases on the earth, so the Jovian planets sometimes are called *gas giants*. However, this is a bit of a misnomer, because much of the interiors of the Jovian planets are, due to the extremely high pressure there, liquid, not gas.

The sixth column lists the rotation period of each planet, again expressed in a sort of terrestrial unit—the day—as defined on earth. Notice that the rotation period of Mars is little more than a day, but that the rotation periods of Mercury and Venus are very long. On the other hand, the rotation periods of all the Jovian planets are very short, in all cases shorter than a day. Therefore, we can conclude that it is common characteristic that Jovian planets rotate quickly, but that terrestrial planets rotate slowly. The seventh column lists the number of known satellites of the planets (as of the time of this writing—new satellites are discovered from time to time so this likely will change). Only two of the four terrestrial planets have satellites, and the two satellites of Mars are very small. Therefore, among the terrestrial planets there are only three satellites, and only one of them is of any appreciable size. However, each of the Jovian planets has many satellites, and many of those are of considerable size.[4] Therefore, in general, terrestrial planets have few or no satellites, while Jovian planets have many satellites. Finally, column eight lists whether a planet has a ring system or not. Notice that none of the terrestrial planets have rings, while all four Jovian planets have rings.

We thus can see that the two types of planets differ in seven ways. How does Pluto stack up in this comparison? In terms of distance from the sun, density, and number of satellites, Pluto clearly is a Jovian planet. However, in terms of size, mass, rotation period, and the rings, Pluto must be a terrestrial planet. Obviously, Pluto does not fit into either category. This is another reason why Pluto ought not to be a planet.

Was There Once a Planet Between Mars and Jupiter?

In the late eighteenth century, there arose an idea that there must be a planet between the orbits of Mars and Jupiter. This idea was spawned by a relationship

---

[4] Four of the Jovian satellites have both larger diameter and more mass than earth's moon: Jupiter's Io, Ganymede, and Callisto, and Saturn's Titan

commonly called Bode's law, though, similar to Grape-Nuts, being neither grapes nor nuts, it was neither Bode's nor a law. It was first proposed in 1766 by the German astronomer Johann Daniel Titius. Two years later, the German astronomer Johann Elert Bode republished, and in the process, popularized the same relationship. The relationship begins with the number 0.4. The next number is generated by adding 0.3 to 0.4, and each succeeding number is obtained by successively doubling 0.3 and adding it to 0.4. The first few terms of this relationship are in the accompanying table, along with the distance of the first six planets from the sun in AU (Table 2). Notice the match between distances "predicted" by this relationship and the actual distances of the planets expressed in AU.

Table 2. Predictions of Bode's Law.

Planet	Bode's Law	Distance from sun (AU)
Mercury	0.4	0.387
Venus	0.7	0.723
Earth	1.0	1.000
Mars	1.6	1.52
Ceres	2.4	2.77
Jupiter	5.2	5.20
Saturn	10.0	9.58
Uranus	19.6	19.3
Neptune		30.2
Pluto	38.8	39.5

Also, notice that this relationship "predicts" the existence of two additional planets, one between Mars and Jupiter, and the other about twice the distance of Saturn from the sun. In 1781, a little more than a decade after Bode's publication, the German-born English astronomer William Herschel discovered Uranus, close to the position where Bode's law indicated there ought to be a planet. This circumstance confirmed Bode's law in the minds of some people, fueling expectation that there was another planet between Mars and Jupiter yet to be discovered. In 1801, when the Italian astronomer Giuseppe Piazzi discovered the first minor planet, Ceres, orbiting in the location "predicted" by Bode's law, Bode's law seemed firmly established. Nearly a half century later, the discovery of Neptune clearly disproved Bode's law: Bode's law predicted the position of the next planet at 39.2 AU from the sun, but Neptune is only 30.1 AU from the sun. Interestingly, the predicted distance fits Pluto's distance from the sun (39.5 AU) about as well as it does the other planets, but given Pluto's quirky status just discussed, it appears to be the interloper, not Neptune.

The discovery of Neptune ought to have sufficed to discredit Bode's law, but since it was widely believed for more than a half century, its influence has been difficult to overcome. Even today there are some who cling to the expectation that there ought to be some numerical relationship between planetary distances from the sun. This is particularly strong among a few creationists who insist that there must be some design feature at work, and thus labor to develop mathematical relationships similar to Bode's law but that better fit the data. What is wrong with Bode's law, other than the fact that it utterly fails in predicting planetary distances beyond Saturn? First, it is not a proper mathematical series. The first two terms are arbitrarily set, after which successive terms follow. However, this is not how mathematical series work. There ought to be a simple formula as a function of n that dictates each term, with an indexing that begins with n equal to either zero or one and incrementing by one to generate each new term. Second, there is no underlying physical mechanism upon which the relationship would follow.[5] This is key, because mathematical series of this type normally follow from the physical principles involved. Many who believe that there is some underlying design feature fail to grasp this important element.

Another problem with Bode's law is that it supposedly predicted a planet between Mars and Jupiter, but, despite being in the right place, Ceres hardly is large enough to qualify as a planet. The discovery of other minor planets with nearly the same orbital distance from the sun as Ceres soon followed. Eventually, astronomers came to realize that there were many minor planets between Mars and Jupiter in a region known as the *asteroid belt*. This soon led to speculation that there once was a planet located where the asteroid belt now is, but that planet exploded,[6] leaving the asteroid belt in its wake. Due to

---

[5] The most serious attempts at describing a physical mechanism for Bode's law is orbital resonances, a topic discussed later in this chapter in the context of gaps in Saturn's rings. However, this has not been successful.

[6] This idea quickly became public knowledge, as evidenced by the backstory of Superman, introduced in 1938. Kal-El was the lone survivor of the planet Krypton when it exploded. Kal-El ended up on earth, where he grew to become Superman.

its catastrophic nature, this idea has tremendous appeal to some creationists. It seems that once one comprehends the catastrophic nature of the Flood, it is easy to envision catastrophes of all sorts. One problem with this idea is that planets do not spontaneously explode. A more promising approach is to hypothesize that a planet once existed there, but that a collision with another body disrupted it. Another difficulty, however, with any destroyed planet scenario is that the asteroid belt contains about $1/20$ the moon's mass, or about $1/1600$ the mass of the earth. This hardly would constitute a respectable planet. Despite the appeal that a disrupted planet has for some creationists, it seems that this idea is a solution in search of a problem.

What Do Mars and Venus Reveal?

Biblical creationists believe in design and purpose in the world around us. But just because some aspect of creation has no clearly identified design or purpose does not mean that there is none. Rather, it could mean that we have not yet discovered what design or purpose exists. Such is the situation with many of the planets—we do not know what purpose they serve. However, there are two factors about the planets Venus and Mars that may show purpose.

Astronomers delineate a narrow range of distance from a star where an orbiting planet might have conditions conducive for life called the *habitable zone*. The habitable zone is based primarily on the possibility of liquid water existing on a planet's surface. If a planet orbits too closely to a star, the surface temperature likely will exceed the boiling point of water. On the other hand, if the planet orbits too far from the star, the surface temperature probably will fall below the freezing point of water. This calculation assumes an appropriate atmosphere, both in terms of the amount and composition. It is not surprising that earth falls within the sun's habitable zone. Much research now is devoted to discovering planets orbiting other stars (extrasolar planets) within the habitable zones of those stars.

Both Venus and Mars orbit outside the range of the sun's habitable zone, so they offer the opportunity to demonstrate just how wrong a planet can be when it comes to supporting life. Venus closely matches the earth in size and composition, so much so that at one time astronomers referred to it as earth's twin. However, the similarities end there. The surface temperature of Venus is nearly 900°F, both day and night. Much of the reason for this extremely high temperature is Venus' thick atmosphere that consists almost entirely of carbon dioxide (carbon dioxide is a greenhouse gas, which traps heat). However, many theories about the past on Venus suggest that its hostile atmosphere likely is related to its distance from the sun.

Mars is quite a bit smaller and less massive than the earth, so it has a much thinner atmosphere than the earth. However, we can still make some comparisons. In the tropical regions of Mars, daytime air temperatures briefly can exceed the freezing point of water. However, the air temperature at night frequently plunges to -100°F or lower. Furthermore, over much of the Martian surface the air temperature never exceeds the freezing point and the nighttime temperatures can approach -200°F. Clearly, both Venus and Mars are hostile to life, because they lie too far outside of the sun's habitable zone. God could have directed these harsh conditions on either side of the habitable zone to demonstrate to us the favored status that the earth has. Some people have termed the earth the *Goldilocks planet.*

But Mars and Venus may serve another purpose. Genesis 6–9 records the year-long Flood during the time of Noah. There is abundant evidence that much water once flowed on the surface of Mars. There are numerous stream channels that appear to have been carved by water. Even more importantly, ancient shorelines appear high on Martian hills, suggesting that vast oceans once covered at least part of Mars. Some planetary scientists have proposed that a global or near-global flood once covered Mars, a planet with no liquid water. Yet the same planetary scientists insist that such a thing could not have happened on the earth, a planet literally awash in water. Many physical models of how the Flood may have happened include intense volcanic and/or tectonic activity at the time of the Flood (one example was Catastrophic Plate Tectonics discussed in Chapter 3). The surface of Venus shows evidence that its entire surface was overturned in tectonic upheaval in a relatively short period and not that long ago. Yet planetary scientists largely insist that such a thing could not have happened on earth. It may be that God produced tectonic upheaval on the closest planet and a large Flood on the second closest planet so that man could see that both were possible on the earth.

Mars has two very small satellites, and they have an interesting story to tell. The two were discovered by the American astronomer Asaph Hall in 1877. A century and a half earlier, the satirist Jonathan Swift wrote in *Gulliver's Travels* an account of astronomers of Laputa who had discovered two small satellites of Mars orbiting very close to Mars. The two satellites discovered by Hall also orbit close to Mars, with periods of 7.6 hours and 30.3 hours, respectively. Swift's fictional satellites had periods of 10 hours and 21.5 hours. The uncanny similarities between the real and fictional satellites in terms of number, size, orbital distance, and orbital period have suggested to some people that this is not a coincidence. Some creationists have opined that perhaps the antediluvian civilization had advanced technology that rivaled the technology of today. These creationists further reason that perhaps these ancient technological innovations included large telescopes that enabled them to discover the two small satellites of Mars. Furthermore, they reason that while the technology may have been wiped out by the Flood, some of that ancient knowledge gained may have survived, and perhaps Swift had access to old records of the two Martian satellites—records that now, conveniently, no longer exist. A more bizarre version posits that there were planetary cataclysms in the early solar system, some related to catastrophic events, such as the Flood. One of these supposed cataclysms was a very close approach of Mars to earth—so close that people on earth could see Mars' two satellites with the naked eye.

These make for interesting tales, and one of them often has been used to support the notion of an extremely advanced antediluvian civilization. However, how plausible is this, and is there a simpler, alternate explanation? Or is it just a coincidence? At the time of Swift, it was commonly believed that Mars must have satellites. At the time, there were six known planets. The first two planets from the sun did not appear to have any satellites. The earth very clearly had one satellite. Mars, the next planet, did not appear to have any satellites, but Jupiter had four known satellites, and Saturn had five. To many people, there appeared to be a mathematical progression here.[7] It seemed appropriate, then, that Mars ought to have more than one satellite, but less than four. That left the possibilities of two or three, with two being more likely. However, if Mars had two satellites, how had astronomers failed to discover them already? The

---

[7] This approximately was the time of the popularity of Bode's law, and what appeared to some as a mathematical progression. This mystical intersection of mathematics and the world had other precedents. For instance, Johannes Kepler thought that because there were (at the time) five other planets and five perfect solids, there must be a connection between the two. There are many mathematical progressions that we can relate to nature, but they arise from mathematical expressions of physical principles at work.

only way that they could have been missed is if they were very small and hence very faint. Furthermore, if they orbited very closely to Mars, they were more likely to be lost in the much brighter light of Mars. This was a common belief at the time, and Swift, being a satirist, may have been lampooning those who believed in the existence of these two small satellites of Mars. At any rate, Swift assumed orbital sizes for the two satellites, from which he could compute the orbital periods.[8] But to do this, one must assume the mass of Mars, which was unknown at the time. The most obvious way to do this would be to assume a certain density for Mars. The orbital sizes and periods of Swift's two Martian satellites do not conform to physics, because Swift assumed the wrong density for Mars.

What Do the Outer Planets Reveal?

Three of the four Jovian planets emit about twice as much radiation as they receive from the sun. Uranus is the lone exception. No one knows why Uranus is different in this respect. For that matter, no one knows why the other three Jovian planets emit so much energy. This excess radiation is in the form of infrared (IR) radiation. IR radiation often is called *heat radiation*, because it is the mechanism by which most objects shed heat. Since the clouds in the atmospheres of the Jovian planets are so bright, much of the solar energy that falls on them is reflected. The absorbed energy ought to achieve an energy balance rapidly, so that the IR radiation emitted equals the amount of solar radiation absorbed. Since the output of these three planets greatly exceeds their input, there must be some internal energy source, but what could it be?

The most obvious source of energy would be primordial energy, that is, energy that the Jovian planets formed with. With time, this heat would dissipate. When the planets reached equilibrium, they no longer would radiate excess heat. How long would the planets take to reach equilibrium? It would take a few million years. Now, if the Jovian planets are only thousands of years old, as biblical creationists believe, this is not a problem. But for those who believe that the planets are billions of years old, this is a great problem, because the Jovian planets ought to have reached thermal equilibrium a long time ago. Another possibility is that radioactive elements within the Jovian

---

[8] It has been speculated that Swift received assistance on this calculation by his mathematician friend, John Arbuthnot.

planets could keep them heated. The problem with this solution is that the low densities of the Jovian planets would not permit nearly enough radioactive material to keep them warm for billions of years. Another proposed source of energy is gravitational potential energy. This mechanism would heat the interiors of the Jovian planets as denser material sank to planetary centers and less dense material rose. Again, if this were the mechanism heating the Jovian planets, this source of energy would have been exhausted before the supposed 4.5-billion-year age of the Jovian planets. Since there is no known source for the internal energy of the Jovian planets that would work for billions of years, it would seem to preclude such great age.

A similar thing is true of Io, the innermost large satellite of Jupiter. Io has many active volcanoes on its surface, making Io the most volcanically active object the solar system.[9] How can the interior of Io be so hot, if it is billions of years old? The density of Io is too low to suggest any significant heating from radioactive decay. The usual answer is that Io undergoes tidal flexing as it orbits Jupiter. As the distance between Io and Jupiter changes, the amount of tidal force ought to change. As the tidal force changes, the shape of Io changes back and forth. This flexing can produce heat by converting kinetic energy into thermal energy. However, detailed analysis of the process reveals that the mechanism is not efficient enough to produce the amount of heat required.[10] Io's orbit is very close to being circular, so its distance from Jupiter does not change much. Io's orbit has a very low inclination to the equatorial plane of Jupiter, so Io does not bob up and down significantly with respect to Jupiter's equatorial plane. Perturbing effects of the other three large satellites of Jupiter could play a role, but it appears that their contribution is minimal.

All four Jovian planets have ring systems, though only one, Saturn, has a grand system. Even small telescopes can show Saturn's rings. However, the ring systems of the other three Jovian planets are very faint, and so were not discovered until the 1970s. The rings are made of a huge number of individual, small particles. These particles orbit in the equatorial planes of their planets.

---

[9] Rather than rock, the molten material on Io primarily is sulfur. Hot sulfur also is known as brimstone. Therefore, Io smells like hell.

[10] Wayne R. Spencer, "Tidal Dissipation and the Age of Io," *Proceedings of the Fifth International Conference on Creationism*, edited by R.L Ivey, Jr. (Pittsburgh, Pennsylvania: Creation Science Fellowship, 2003), 585–597

There is a great range in the orbital distances of the particles, so planetary rings are very broad. However, the rings are very thin, perhaps as thin as 100 yd (90 m). Occasionally Saturn passes in front of a bright star. As this happens, we can see the star shining through the rings. This would not happen if the rings were solid, but it would happen if the rings were made of many small particles. Twice each orbital period of Saturn around the sun (that is, twice in nearly 30 years), we pass through the equatorial plane of Saturn. At these times, the rings completely disappear. These things can happen only if the rings are very thin.

The rings of Saturn have very narrow gaps in them. One of the most famous, Cassini's division, can be seen through moderate-sized telescopes. Space probes to Saturn have revealed that there are hundreds of gaps in Saturn's rings. The gaps are caused by gravitational perturbations, or little tugs, from some of Saturn's satellites. For instance, the largest gap, Cassini's division, occurs where there is a 2:1 resonance with Mimas, one of the larger satellites closest to the rings. The orbital period of Mimas is exactly twice that of particles within Cassini's division. Every other orbit the gravity of Mimas will give an extra strong pull on each particle in Cassini's division. For each particle, this pull will be in the same location in its orbit, and it will have the effect of pulling ring particles to higher orbits, effectively cleaning Cassini's division. Astronomers call this a *resonance*. Once a ring particle is out of Cassini's division, it no longer gets a nudge in the same direction every other orbit, so Mimas does not send the particle any higher (Image D). The many other gaps in Saturn's rings have similar explanations with resonances related to other satellites. Any integral ratio of orbital periods will produce resonances, though the lower the integers, the stronger the resonance. However, none of them are nearly as strong as the 2:1 resonance with Mimas.

As many ring particles are pulled to various orbits, they inevitably collide with one another. In addition, meteoroids from outside the rings collide with ring particles. These are much higher energy collisions that the collision of intra-ring particles, usually resulting in removing particles from the rings. Both types of collisions often break the particles into smaller pieces. The ring particles are made of ice and dust. The collisions frequently break the chemical bonds holding the ice molecules together, forming ions. Solar wind and the magnetic field of Saturn can further accelerate the ions, removing them from the rings. These forces and others gradually wear down the rings. How long can a ring system last? Estimates are that the ring systems can last only a few million years. Again, this is not a problem, if the rings are only thousands of years old; but how can ring systems persist if the solar system is billions of years old?

One theory is that ring systems can form from the shattering of a satellite. There is a limit to how close to a planet a satellite that is held together by gravity may approach. If a satellite's orbit is altered so that it comes within this *Roche limit*, then the tidal force of the planet will overcome the satellite's gravity and tear the satellite apart. This is how many astronomers think that rings formed. Keep in mind that this limit applies only to satellites that are held together by their gravity. Small satellites are held together by normal chemical bonds, so the Roche limit does not apply to them. Would this mechanism for ring formation explain why there are ring systems in a solar system that is billions of years old? It could, but we must keep several things in mind. First, this would mean that all four planetary rings now present in the solar system formed only recently and hence are young. However, what is the probability that we just happen to live at a time in which all four Jovian planets have ring systems? Second the Jovian planets have a limited supply of satellites large enough to produce rings. Just how many times in 4.5 billion years have the Jovian planets formed rings and then quickly lost them? Given that satellites generally have reasonably stable orbits around their planets, what is the mechanism that so altered their orbits?

What Do the Small Bodies of the Solar System Reveal?

A comet consists of a small nucleus only a few miles across. The nucleus is made of various ices, such as water ice, dry ice, and frozen methane and ammonia. Mixed in with the ices are microscopic dust particles. The densities of comet nuclei appear to be lower than the densities of the ices and dust, implying that comet nuclei are very porous. With their long, elliptical orbits, comet nuclei spend most of the time far from the sun, where the temperature is low enough for the ices to remain frozen. However, once each orbit a comet nucleus approaches the sun. When close to the sun, the nucleus absorbs enough solar radiation to melt some of the ices. Being in space with essentially no pressure, the ices sublime immediately into gas without passing through the liquid phase. The gas rapidly expands to form a large coma around the nucleus. The coma may be thousands of miles across. Solar radiation ionizes the gas, and as electrons recombine, they emit light. This process can make the coma appear very bright. Solar wind particles push the ions away from the sun to form the ion tail. The sublimation of ices also dislodges dust particles. The dust particles shine by reflecting sunlight. Sunlight also exerts a force on dust particles, pushing them outward to form the dust tail. A comet brightens tremendously when near perihelion, but soon fades into obscurity as the comet passes away from the inner solar system.

Clearly, a comet nucleus loses a tremendous amount of material each time it passes near the sun. Given that comet nuclei are not large to begin with, there must be a finite lifetime for comets. Astronomers estimate that a typical comet nucleus could not survive more than a hundred close encounters with the sun. Many comets have long orbital periods, but there is a limit to how long their orbital periods can be. If a comet's aphelion was a significant distance toward the nearest stars, its period would be on the order of a few million years. If a comet's aphelion were much farther, it could not be a permanent member of the solar system. If we multiply the maximum orbital period by the maximum number of orbits that a comet could have survived a close perihelion passage, we conclude that comets could not have orbited the sun for more than a few hundred million years. That is, if the solar system were billions of years old, then why are there any comets left?

There are other mechanisms that destroy comets as well. Occasionally a comet collides with a planet. This was known to be a possibility, but it was not observed until 1994, when Comet Shoemaker-Levy IX collided with Jupiter. Jupiter survived this event relatively unscathed,[11] but the comet met a sudden and catastrophic end. But a comet need not collide with a planet to be eliminated. If a comet passes close enough to a planet, the planet's gravity can dramatically alter the comet's period. About half the time, Jupiter's gravity can add energy to a comet's orbit. If a comet is barely bound to the solar system, which frequently is the case, the addition of orbital energy will eject the comet from the solar system, never to return. This too represents a catastrophic loss

---

[11] Jupiter's strong gravity had shredded the comet into pieces during a close pass of Jupiter two years earlier on its way toward perihelion. After perihelion passage, the comet's orbit carried it on a collision course with Jupiter. The pieces crashed into Jupiter over several days, leaving dark welts in Jupiter's atmosphere, but those blemishes soon vanished.

of comets, and this has been observed several times. On the other hand, about half the time, the gravity of a planet decreases the orbital period of a comet. With more frequent passes to perihelion and the vicinity of the planets, the other two loss mechanisms work even more efficiently to remove comets.

Collectively, these loss mechanisms are efficient enough that there ought not be any comets after a few hundred million years, let alone billions of years. Astronomers who believe in billions of years are aware of this problem, so they have hypothesized a source of comets that replaces comets as they are lost. One supposed source of comets is the Oort cloud, hypothesized by the Dutch Astronomer Jan Oort around 1950. He suggested that there is a large reservoir of comet nuclei orbiting far from the sun in a roughly spherical distribution. At such great distance, the comet nuclei would be too small to be detected. Furthermore, their equilibrium temperature would be so low that they would remain frozen indefinitely. Oort thought that the gravity of an occasional passing star or other agents outside the solar system would alter the orbits of objects in the Oort cloud so that new comets would plunge into the inner solar system to replace older comets as they were eliminated.

For more than three decades, astronomers thought that the Oort cloud could explain the existence of all comets in this manner. However, studies during the 1980s called this into question. Clearly there are two types of comets: long period and short period, with a dividing line in orbital period of approximately 200 years. However, the difference between the two types is more fundamental that just the lengths of the orbital periods involved. Short-period comets tend to have lower orbital inclinations than long-period comets, and they tend to orbit the sun in the same direction that the planets orbit the sun (prograde). Long-period comets can have any orbital inclination, and about half orbit prograde, while the other half orbit the sun retrograde (opposite the direction that the planets orbit the sun). There does not appear to be a difference in composition between the two types of comets. For years, astronomers thought that planetary perturbations, primarily those caused by Jupiter, could convert long-period comets into short-period comets. However, simulations in the 1980s showed that the efficiency of that mechanism was woefully inadequate to account for the number of short-period comets.

In response, astronomers resurrected the Kuiper belt, a concept proposed by the American astronomer Gerard Kuiper about the same time that Oort published his work on comets. The Kuiper belt had not received much attention, because it supposedly existed early in the solar system, but had long since dissipated. Based upon an evolutionary theory about the formation of the planets, Kuiper thought that early in the solar system, many small objects orbited the sun beyond the orbit of Neptune. These supposedly were bodies that failed to amalgamate into a planet. Kuiper thought that this belt long ago disappeared, because the gravitational perturbations of the outer planets would have eliminated objects in it. Presently, the thinking is that many of the objects originally in the Kuiper belt are still there, and the gravitational perturbations of the outer planets serve to inject new short-period comets into the inner solar system as older comets are eliminated. Additionally, the current theory is that early in the solar system's history, gravitational perturbations of the outer planets populated the Oort cloud from the Kuiper belt. New long-period comets merely are returning close to their birth places.

Is there any evidence that any of this is true? If the Oort cloud exists, objects the size of comet nuclei within the Oort cloud are much too faint to be visible from earth, so there is no evidence that the Oort cloud exists. Starting in the 1990s, the diligent search for small bodies orbiting the sun beyond Neptune's orbit began. Since then, many such trans-Neptunian objects (TNOs) have been discovered. Since this is the region where the Kuiper belt is predicted to have been, this has been hailed as confirmation of the Kuiper belt. However, there are problems with this interpretation. Many of the objects thus far discovered are orders of magnitude larger than any comet nucleus ever observed. One must wonder why we have never seen such large comets. Another problem is that we know the density of a few objects in the Kuiper belt, such as Pluto and Eris. The densities are too high to account for the known composition of comets, so at least the large objects among the TNOs are not a good match for comets.

Speaking of Pluto, it also presents problems for an ancient solar system. During the summer of 2015 the New Horizons mission flew past Pluto and sent back incredible photos of Pluto and its largest satellite, Charon. Being small bodies on the edge of the solar system, virtually all astronomers thought that Pluto and Charon had not experienced any significant geological activity for a very long time. Consequently, astronomers expected their surfaces to be among the most heavily cratered surfaces in the solar system. Both surfaces have craters, but not nearly as many as expected. Some regions of Pluto are nearly void of craters. This is possible only if both bodies have experienced geological processes relatively recently. But such processes must be driven by heat, so what could be the source of energy? As in previous discussions with regard to the internal heat of Io and Jupiter, astronomers have three possible sources at their disposal.

First, there is primordial heat. However, if Pluto and Charon are billions of years old, they would have long ago shed all their primordial heat. The second possibility is radioactive decay of elements within. This is what is the supposed source of the earth's internal heat. But long half-life radioactive elements are very heavy, and the density of Pluto and Charon are far too low for there to be any significant radioactive isotopes present. The third possibility is tidal flexing, as is invoked to explain Io's internal heat. But Pluto and Charon lack the mass to raise tides on each other strong enough to produce sufficient tidal heating, and there are no nearby bodies massive enough to accomplish this. Of course, if Pluto and Charon are young, as in the biblical creation model, then there is no problem to explain—both Pluto and Charon could have had geological activity in the past spawned by primordial heat.

Where Did the Solar System Come From?

In the creationary model, the answer to this question is easy enough—God made objects in the solar system on Day Four when He made all astronomical bodies. But the evolutionary model is committed to naturalistic explanations only. According to the evolutionary theory, the solar system formed from the collapse of a gas cloud a little more than 4.5 billion years ago. This hypothetical process is directly related to the theory of the naturalistic formation of stars, a topic that I shall discuss in a later chapter. Suffice it to say for now that gas clouds do not spontaneously contract to form stars. Assuming that this collapse somehow happened, the theory dictates that most of the matter comprising the gas cloud fell toward the center to form the sun. Most of the remaining material flattened into a disk. This disk represents the plane of the solar system in which the planets' orbits are found. Small particles in the flattened disk began to stick together to form larger particles, though the reason why this happened is not understood either. Astronomers call these amalgamating particles *planetesimals*. With time, the planetesimals grew by assimilating other particles. Eventually, some particles became massive enough that their gravity could attract additional particles to them, greatly speeding up the process of accretion. Some of the planetesimals grew very large, resulting in the formation of planets. Other particles did not get nearly as large. They became satellites, minor planets, or comets.

Location supposedly played a key role in what sort of bodies formed. As the proto-sun developed, its radiation heated the inner solar system so that elements and compounds with low boiling points were evaporated and blown outward from the sun. The bulk of the parent gas cloud was hydrogen and helium, elements with low boiling points, so most of the hydrogen and helium was blown away from the inner solar system. The remaining materials made up at most only a few percent of the original gas cloud, so relatively little material remained in the inner solar system. And that remaining material tended to be solid and dense. This is the evolutionary explanation of why the inner solar system planets (the terrestrial planets) are so small, rocky, and dense. On the other hand, the sun's radiation was too weak in the outer system to remove the low boiling-point materials there, so much of the hydrogen and helium remained there. This supposedly explains why the outer planets are so massive, yet have low density. The transition zone between the two distinct compositions is near the location of the asteroid belt. Thus, SSSBs tend to fall into one of two groups—minor planets or comets—depending upon where they formed.

For some reason, material in two regions of the solar system failed to accumulate into a planet. One region is where the asteroid belt is located, between the orbits of Mars and Jupiter. Presumably, Jupiter formed relatively early, and the perturbing effect of its gravity kept the material in the asteroid belt stirred up, preventing it from forming into a planet. The second region where a planet failed to form is the region of the TNOs, or, as some prefer to call it, the Kuiper belt. Again, the rapid formation and perturbing gravity of the massive Jovian planets have been invoked as the agent that kept material in this region sufficiently churned to prevent a planet forming there. Furthermore, the continued perturbations of the gravity of the outer planets supposedly have populated the Oort cloud from which long-period comets come and introduce new-short period comets directly into the inner solar system.

While this theory may explain why there are two distinct types of planets in the solar system, it fails to explain other features. For instance, it does not explain why the Jovian planets rotate so quickly and the terrestrial planets generally rotate so slowly. Nor does it explain why two planets rotate backwards (Venus and Uranus), or why planets have all sorts of tilts to their axes. Half of the planets have moderate tilts. These include earth (23.4°), Mars (25.2°), Saturn (26.7°), and Neptune (29.6°). On the other hand, three planets have little tilt: Mercury (0.0°), Venus (2.7°, though this normally is expressed as 177.3°, since it rotates retrograde), and Jupiter (3.1°). Then there is Uranus. Its axis is tilted 82.1°, though this usually is stated as 97.9°, since it rotates retrograde. Either way, the tilt of Uranus' axis is nearly perpendicular to its orbital axis. To explain these oddities, evolutionary astronomers appeal to collisions with large bodies that planets suffered late in their formation. For instance, if a collision is properly oriented, it can act as a torque to slow and even reverse a planet's spin. A different orientation of a large impact could change the tilt of the rotation axis.

This introduces a new problem with the major satellites of Jupiter, Saturn, and Uranus. The large satellites of these three planets orbit in the equatorial planes of their respective planets (the orbits typically are tilted by a degree or less) and in the same direction of rotation as the respective planets. Evolutionary astronomers think that each of these planets and their satellites formed from a common reservoir. As most of the matter in the solar system sank to the center to form the sun, the remaining material flattened into a disk from which the planets formed. Most of the matter in this disk then coalesced to form the planets, with the remaining material subsequently flattened (again) into smaller disks from which the planets' satellites formed. If this idea is correct, then the collisions that reoriented the axes of the planets occurred *after* the segregation of material that formed the planets and satellites. Thus, the material that formed the satellites did not share in this reorientation. Supposedly, gravitational forces of the massive planets in conjunction with other effects gradually reoriented the orbital planes of satellites to the new equatorial planes of the respective planets. This is a tricky proposition physically due to angular momentum considerations, but perhaps it could work. However, this explanation is problematic with Uranus. Not only is Uranus highly tilted, but it rotates backwards. Its satellites share in this backward motion, but this process of reorientation cannot reverse their direction. The answer to this problem is to invoke *two* major reorienting impacts of Uranus, with sufficient time between the two events to reorient the orbital planes of the satellites.

You may have noticed that the other Jovian planet, Neptune, and its satellites, are conspicuously absent from this discussion. Neptune has two major satellites. The larger one, Triton, has an orbital inclination of 157.3°. This means that Triton orbits Neptune backwards on a moderately inclined orbit (22.7°, if expressed as retrograde). Neptune's smaller major satellite, Nereid, orbits in the proper direction, but with an inclination of 32.6°. Obviously, the scenario laid out above does not suffice to explain these two satellites of Neptune. The usual explanation for these two satellites is that they did not form with Neptune but instead were captured by Neptune's gravity. Indeed, this process is invoked to explain some of the small satellites of the other Jovian planets. With so many SSSBs, there certainly is a large reservoir of potential bodies to capture. The capture of a satellite is not a simple matter. If a massive body (call it the first body) is approached by a small second body, the gravity of the first body will alter the trajectory of the second body. However, that alteration is insufficient to capture the second body into an orbit around the first body. Capture requires the interaction of at least one additional body. The complex interaction of this third body with the other two can rob the second body of enough kinetic energy to cause the second body to orbit the first body. The third body could be another random object like the second body, with both just happening to pass close to the first body. Or the third body could be an object already orbiting the first body (this would make it a satellite already). The kinetic energy lost by the second body is gained by the third body. If the third body is a satellite, then its orbit will be altered, either moving it to a higher orbit, or stripping it from the planet altogether.

Capture is a random process, so the orbital characteristics of a captured body ought to be random. That is, we would expect satellites captured by planets to have inclined, eccentric orbits. Indeed, there are many small satellites of the Jovian planets whose orbits match this description. In addition to being very small, they orbit far from their respective planets. Most orbit retrograde. Astronomers call them *irregular satellites*. How do Triton and Nereid compare to the irregular satellites? Like other irregular satellites, they orbit far from their planet, and they have high orbital inclinations (and Triton orbits retrograde). Nereid has a highly eccentric orbit, which is characteristic of irregular satellites. However, Triton has a very circular orbit (among the larger satellites, only Saturn's Tethys has a more circular orbit). This is very different from typical irregular satellites. Finally, Triton and Nereid are far larger than any of the other irregular satellites. Oddly, the smaller satellites of Neptune are boringly regular, with close, low inclination, low eccentricity orbits. Most baffling is Triton, whose inclined, retrograde orbit virtually screams "capture!" while its very low eccentricity strongly argues against capture. Putting this all together, the satellites of Neptune defy simple explanation from a naturalistic standpoint. As with the Uranian system, Neptune requires a series of highly unlikely events.

The theory of the naturalistic origin of the solar system has faced other challenges. The sun contains 99% or more of the mass of the solar system, yet it contains only about 1% of the angular momentum. Angular momentum ought to be more equitably distributed in proportion to mass, so biblical creationists often have made use of this disparity in criticizing the evolutionary theory of the origin of the solar system. The standard answer is that the magnetic field of the forming solar system transferred angular momentum from the proto sun to the disk of the forming solar system. Such a mechanism indeed might have worked to accomplish this, so this may not be a legitimate argument against the evolutionary scenario for the solar system's origin. This topic deserves more study in the creation literature.

It is noteworthy that this theory of the origin of the solar system was worked out before the discovery of extrasolar planets, planets orbiting other stars. This theory results in large, gaseous planets orbiting far from the sun, with small, rocky planets orbiting much closer to the sun. In fact, this theory *cannot* account for large, gaseous planets near a star. However, we now know that large, and presumably gaseous, planets that orbit close to their stars are common. To resolve this contradiction, theorists now claim that planet migration is common, at least early in the history of planetary systems. As with the proposed notion of how irregular satellites are captured, the interaction of planets supposedly can cause some planets to migrate inwards while others migrate outward. So the large planets that we detect orbiting closely to their stars originally started out much farther away. Theorists have worked out scenarios of how planets may have migrated early in the history of the solar system. The takeaway is that the solar system may be unique, or at the very least, unusual, as planetary systems go. Scientists thought that the discovery of extrasolar planets would confirm that their theory of the formation of planets within the solar system was correct. However, the discovery of planets orbiting other stars had the opposite effect. Consequently, scientists had to modify greatly their theories to resolve this issue. But how can scientists be so sure that this time they got it right?

In the previous chapter, I discussed the unique orbit of the moon, and the unique origin scenario for the moon that evolutionary scientists are forced to invoke to explain it. In this brief survey of the evolutionary ideas of the origin and history of the solar system, I have described many unusual and even catastrophic events that a naturalistic origin of the solar system requires. The fact that so many ad hoc adjustments must be made demonstrates that the belief in a naturalistic origin for the solar system fails. These adjustments amount to "just-so" stories, in that while they *might* have happened, there is no evidence that they actually *did* happen. Many of these events are highly improbable, which ought to indicate a problem. Furthermore, nearly all—if not all—of these cases of special pleadings call for catastrophic events—events far more catastrophic than the Genesis Flood. Yet supporters of the naturalistic theory of the solar system's origin would dismiss the biblical Flood out of hand.

CHAPTER 6

The Greater Light to Rule the Day: The Sun as a Star

Today we understand that the sun is a star. Thus, the study of the sun can prepare the way for our study of other stars. People have not always thought of the sun as a star. Indeed, the sun certainly appears very different from the stars. Obviously, the sun appears much brighter than the stars. Furthermore, we can see that the sun has a physical diameter, but stars appear as mere points of light, even when viewed through the largest telescopes. We explain the obvious differences between the sun and stars by the fact that the stars are much farther away than the sun. The nearest star is 275,000 times farther away than the sun is, and most stars that we see in the sky are millions of times farther away. This does not mean that thinking that the sun is not a star is wrong, or that thinking the sun is a star necessarily is right. It just means that these are two different ways of thinking about the sun. A good case could be made for either, based upon what criteria we happen to think are more important. Indeed, in common usage today the sun is distinctly different from the stars, even though nearly everyone agrees that the sun is a star. Notably, the Bible never refers to the sun as a star. That allows for the possibility that perhaps the sun is fundamentally different from other stars.

The sun is a sphere of hot gas about 860,000 mi (1,400,000 km) in diameter. That is 109 times the diameter of earth. The sun's mass is a third of a million times that of the earth. About 75% of the sun's mass is hydrogen. Much of the remainder is helium. Less than 2% of the sun is made of all the other elements. The sun is very hot; the surface temperature is 5770 K, nearly 10,000°F. As you might expect, the temperature, as well as density and pressure, increase with depth inside the sun. While we cannot directly measure the conditions in the interior of the sun, we can estimate them there. The temperature in the sun's core is more than 15,000,000 K, nearly 28,000,000°F. The pressure in the sun's core is about 250 billion times that of the pressure on the surface of the earth. The density in the solar core is about 150 grams per cubic centimeter. This is more than ten times the density of lead, yet the material is a gas, because of the extreme pressure and temperature there. The temperature, pressure, and density gradually decrease from the sun's core to its surface. Throughout most of the sun's interior, the temperature is too high for atoms to exist. Instead, the matter is ionized, with a mix of atomic nuclei and free electrons. Only near the solar surface can atoms exist, but even there most of atoms are at least partially ionized.

Three Layers of the Sun's Atmosphere

A word of warning about viewing the sun: the sun is extremely bright, and exposing one's eyes to the full light of the sun even briefly can result in blindness. There are a few ways safely to view the sun. However, if you do not know what you are doing, please do not attempt this. The safest way to view the sun is to place a special solar filter on the front of a telescope. These filters look like mirrors, because they are designed to reflect almost all the sun's light. Only a small amount of the sun's light passes through the filters. This transmission of light is nearly uniform across the entire visible spectrum. When light of all colors is combined, white light results, so we call these solar filters white-light filters. Despite the name, the sun may appear a shade of yellow, because that is the true color of the sun.

When viewed safely in this manner, we are looking at the sun's *photosphere*, the bottom-most part of the sun's atmosphere, from which comes nearly all the sun's light (Image E). The word *photosphere* comes from two Greek words, meaning "light sphere." The sun's limb, or edge, may appear a little dimmer than the center of the sun's image. We call this *limb darkening*. What causes limb darkening? When we look at the center of the sun's image, we are looking straight down into the sun's photosphere, but when we look near the sun's limb, we are looking at nearly a grazing angle to the photosphere. Therefore, we do not see as deeply into the photosphere near the sun's limb as we do at the center of the sun's image. Temperature decreases with height in the photosphere, so the level of the photosphere that we see near the limb is cooler than near the center of the sun's image. Cooler objects appear fainter than brighter ones, so the sun's limb appears darker than the rest of the sun.

The photosphere is a few hundred miles deep. Above the photosphere there are two layers of the sun's atmosphere. Directly above the photosphere is the *chromosphere*, from two Greek words meaning "color sphere" (Image F). Astronomers chose this name because the chromosphere was first detected by the very brief, colorful spectrum that it produces at the beginning and ending of a total solar eclipse. The chromosphere is much fainter than the photosphere, so normally it is visible only during total solar eclipses. The chromosphere shines by emitting light in a few very narrow portions of the spectrum in what we call *emission lines*. The photosphere absorbs energy at the same narrow wavelengths that the chromosphere emits energy, producing *absorption lines*. Since the chromosphere emits energy at the same wavelengths that the photosphere absorbs energy, the chromosphere appears relatively bright compared to the photosphere when viewed at these narrow wavelengths. Therefore, astronomers can study the chromosphere anytime by using special filters that transmit light only in these narrow ranges of the spectrum. One of those special filters is an H-alpha filter, called so because its transmission is centered on an emission/absorption line of hydrogen in the red part of the spectrum.[1] Solar prominences, mentioned in Chapter 4 in conjunction with total solar eclipses, emit most of their energy in H-alpha, so they are easily seen in an H-alpha filter. As in the photosphere, in the lower chromosphere the temperature continues to fall with height, but higher in the chromosphere, this trend reverses, and the temperature increases with higher elevation.

Above the chromosphere is the *corona*, the Latin word for "crown." The pearly-white corona extends outward a few solar radii, so the corona is the thickest part of the sun's atmosphere by far. While some special instruments can reveal the outer corona, the corona is best observed during total solar eclipses. The temperature increase with altitude continues in the corona, with temperature reaching into the millions of Kelvin. While this is an incredibly high temperature, there is not a lot of heat in the corona. This may sound paradoxical, but it underscores the difference between temperature and heat.

---

[1] The alpha designation comes from this line being the first in a series of lines called the Balmer series.

We all learned in school that heat is related to the motion of particles. More specifically, when particles move, they have *kinetic energy*, or energy that is due to their motion. The equation for kinetic energy is $\frac{1}{2}mv^2$, where m is the mass of a particle, and v is its velocity. In the case of heat, the particles involved are the atoms or molecules making up the object. The atoms or molecules have different velocities, so they have different kinetic energies. The temperature is defined as a measure of the average kinetic energy of the particles. As heat is added, the particles move faster, increasing the average kinetic energy, and so the temperature increases. The heat is the sum of the kinetic energies of all the particles, so the heat could be defined as the product of the average kinetic energy and the number of particles. Normally, there are many particles involved, so high temperature typically is associated with a large amount of heat. However, the density of the corona is very low, so that any volume, such as a cubic meter, contains very few particles. Therefore, the heat, being the product of the average kinetic energy per particle and the number of particles, is not very great, despite the particles having a very high temperature.

The extremely high temperature of the corona presents astronomers with the *coronal heating problem*. From the second law of thermodynamics, we know that heat tends to flow from hotter objects to cooler objects. The chromosphere and photosphere below the corona are much cooler, as are regions beyond the corona—so why does the corona maintain such a high temperature? Some creationists have argued that the solar corona's high temperature indicates that the sun is very young. Presumably, God made the solar corona very hot, and the corona has not had sufficient time to cool in just thousands of years, but it would be cool if the sun were billions of years old. The cooling time of the corona likely is much less than thousands of years, so this probably is not a good argument for the sun's youth. There is no agreement as to the source of the corona's heat, but there are several competing theories, such as acoustic waves from the photosphere and magnetic field reconnection.

The Sunspot Cycle

When safely viewed with even a small telescope, the sun's surface often manifests sunspots. Sunspots appear dark compared to the rest of the photosphere, but this is an illusion. Sunspots are very hot, but just not as hot as the rest of the photosphere, so they appear dark. The inner portion of sunspots often appear darker than their outer regions. Astronomers call the darker inner part of a spot the umbra and the lighter outer part the penumbra.[2] Sunspots often appear in pairs, or in small groups dominated by a large pair. Sunspots last a few days or weeks, so day by day, we can see changes in them. Part of the daily change is due to the sun's rotation. It takes about a month for the sun to rotate once, so spots near the center of the sun's image noticeably change position from one day to the next. Spots normally are within 25° latitude of the sun's equator. The number of spots on the sun varies over a period that averages about 11.2 years. During the year or two of sunspot maximum, hardly a day goes by without a few spots. However, during the years around sunspot minimum, weeks or even months can pass without any spots.

What are sunspots? Besides being cooler regions, they are regions of strong magnetic fields. Unlike the earth, which has a strong global magnetic field with well-defined north and south poles, the sun has a very weak global magnetic field. The magnetic field of the sun is dominated by the sum of strong local fields. We measure magnetic fields in astronomical bodies with the *Zeeman effect*. The Zeeman effect is the splitting of spectral lines into two or more lines in the presence of a magnetic field. Within a sunspot pair, one spot will have north magnetic polarity, while its mate will have the opposite magnetic polarity. On one side of the sun's equator, all the sunspot pairs will be oriented with their north-south polarity in the same direction parallel to the equator, but in the other hemisphere, the polarity will be reversed. This orientation will remain throughout one sunspot cycle from minimum through maximum back to the minimum. However, in the next sunspot cycle, the orientation will reverse, before flipping back to the original orientation the next cycle. Therefore, the sunspot cycle more accurately is 22 years, not 11 years. The first sunspots of a new cycle generally are at higher latitude, around 25°. As the cycle progresses, sunspots occur at decreasing latitude until the cycle ends with sunspots at around 5° latitude. It is common for the last few spots of an ending cycle to overlap with the first few spots of a new cycle, separated by about 20° of latitude.

For more than a half century, the Babcock theory has done a good job of explaining sunspots (Figure 6.1). It proposes that a sunspot cycle begins with a weak global magnetic field on the sun. Being an ionized sphere of gas

---

[2] Note that astronomers have recycled these two terms from the parts of shadows relevant to eclipses.

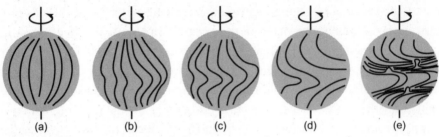

Figure 6.1. Illustration of Babcock theory.

with a magnetic field threading through it, the sun is a plasma. Since the sun is gaseous throughout, it does not rotate with a single rate. Rather, the sun rotates differentially, with the equator taking about 25 days to spin and the poles more than 30 days. The differential rotation winds the magnetic field lines, eventually resulting in crowding of the magnetic field at some locations. The crowded field lines indicate a strong local magnetic field. The field lines expand to lessen the local field, but being a plasma, the gas is carried with the field lines. This leads to expansion of the gas, resulting in the cooling and buoyancy of the gas. Buoyant force moves the gas upward, taking the magnetic field with it. The field erupts through the surface in two places. The two places where the magnetic field passes through the solar surface are cooler than the surrounding photosphere, resulting in sunspots. They also have opposite magnetic polarities, with all such pairs having the same orientation in one hemisphere, but having the opposite polarity in the other hemisphere. As a sunspot cycle continues, the magnetic field lines disconnect and reconnect, weakening the field. Eventually, a new, reversed magnetic field generates, and the cycle begins again. While this theory explains much of the phenomena surrounding the sunspot cycle, it fails to address what generates the sun's weak global field to begin with.

There is a strong correlation between the amount of sunspot activity and climate on the earth. This correlation is not over the 11-year cycle, but rather over several cycles. When sunspot activity remains strong over several cycles (many spots during maximum), there tend to be warmer temperatures on large portions of earth, and when sunspot activity is diminished over several cycles, there tend to be cooler temperatures. Most notable in this correlation is the Maunder minimum, a 70-year period in the seventeenth and early eighteenth centuries when there were very few sunspots. This coincided with the *little ice age*, when temperatures in Europe and North America were much colder than usual. The *Medieval Warm Period*, when temperatures in the north Atlantic, Europe, and North America were exceptionally warm, existed from about AD 950–1250. Notably, during this time Vikings from Scandinavia settled the southwest coast of Greenland, where they raised crops and grazed cattle for generations, things that cannot be done today because it is far too cold. From indirect evidence, scientists have deduced that solar activity during this period was extremely high too. However, mere correlation does not necessarily imply causation. No one has yet identified a mechanism whereby the amount of sunspot activity might *affect* climate on earth.

Since 1980, many scientists have become increasingly concerned with a rise in global temperatures on earth. This often is attributed to increased greenhouse gases placed into the earth's atmosphere by human activities, such as burning fossil fuels at unprecedented rates. This has resulted in calls for drastic measures to reduce greenhouse emissions. However, since 1980, sunspot activity has been at near-record highs. If a significant portion of warmer temperatures is due to solar activity, then these actions to reduce greenhouse emissions may not be warranted.

Related to sunspot activity is the *solar wind*. The solar wind is an outrush of charged particles—mostly protons and nuclei of helium atoms—from the sun. Apparently, this behavior is common among stars, for astronomers have detected stellar winds emanating from some stars as well. Solar wind particles move very quickly, only requiring a day or two to traverse the 93,000,000 mi (150,000,000 km) separating the earth from the sun. When the particles reach the earth, the earth's magnetic field deflects those particles. Some of the deflected particles are captured, forcing them to orbit the earth in bands, a region that we call the *Van Allen Radiation Belts*. This deflection is important, because without it, solar wind particles would gradually strip away the earth's atmosphere. Near the earth's magnetic poles, solar wind particles draw closest to the earth, and collisions of the particles with atoms in the air about 60 mi (100 km) up ionize the gas. As the electrons recombine with the atoms, they emit light. This is an *aurora*. In the Northern Hemisphere, aurorae colloquially are called the *northern lights*. In the Southern Hemisphere, they are called the *southern lights*.

Aurorae are most often seen at high latitudes, near the magnetic poles. However, during sunspot maximum, *solar flares* are common. A solar flare is a brief, violent eruption on the sun's surface. Solar flares are hardly noticeable to the eye, because most of the solar flare energy is emitted in parts of the spectrum that we cannot see. Solar flares also eject many charged particles, amounting to a gust in the solar wind. If the gust is directed toward the earth, when it arrives at the earth a day or two later, the earth's magnetic field is a bit overwhelmed. We call this a magnetic storm. During a magnetic storm, some radio and other communications can be interrupted. Electrical transmission systems can be damaged. But the most common effect of a magnetic storm is that aurorae can be seen at much lower latitudes than normal.

What Powers the Sun?

The sun radiates an incredible 3.8×10^{26} watts. Therefore, each second the sun must produce 3.8×10^{26} joules of energy. Where does this energy come from? For much of history, people did not worry much about such a question. However, by the end of the eighteenth century, physicists realized that energy does not simply materialize out of nothing, but instead must come from some source. Scientists first suggested that perhaps some fuel, such as coal, was consumed. However, such large combustion would require a huge amount of oxygen. Where was this oxygen? Furthermore, combustion of a fuel produces waste products, such as ash, soot, and carbon dioxide. Eventually, this waste would build up to the point that it would stifle further combustion. Clearly this is not happening.

By the second half of the nineteenth century, two physicists proposed the first workable theory of the sun's energy. The German physicist Hermann von Helmholtz suggested that the sun slowly shrank, gradually releasing gravitational potential energy. This is the same principle that hydroelectric dams use to generate electricity. Water above the dam has more gravitational potential energy than water below the dam. Water falls from above the dam to below the dam, converting gravitational potential energy into kinetic energy (energy of motion). As the rapidly moving water passes through a turbine, the water slows down, transferring its kinetic energy to a spinning turbine. The turbine spins a generator, which converts the kinetic energy into electricity.

The Scottish physicist, William Thomson, 1st Baron Kelvin, took Helmholtz's work further. Lord Kelvin showed that this mechanism could have powered the sun for no more than about 30 million years. The sun's age could be less than this, but it could be no older, if powered by this mechanism. As discussed in Chapter 3, by the end of the nineteenth century, biologists and geologists thought that the world must be at least 100 million years old, based upon their estimates of how much time was required for geological and biological evolution. Rather than reevaluate their commitment to evolution, most of these scientists assumed that the sun must have another source of energy, though that source remained unidentified for decades. Therefore, astronomers came to abandon what we now call the Kelvin-Helmholtz mechanism in favor of some yet unknown source for the sun's energy. This was not because the Kelvin-Helmholtz mechanism does not work, for indeed it does. Even within the standard evolutionary model of the sun, astronomers believe that early in its history, the Kelvin-Helmholtz mechanism powered the sun, as well as all stars in their early stages. Rather, the abandonment of the Kelvin-Helmholtz mechanism came from the fact that it failed to operate long enough in the estimation of most scientists.

As nuclear physics developed in the twentieth century, astronomers began to consider the possibility that the sun could be nuclear powered. This theory was fully developed by the 1950s. The extremely high temperature and density of the sun's core are sufficient to cause nuclei of hydrogen atoms to fuse into helium, releasing energy. We know how much energy this reaction produces. Since the sun consists mostly of hydrogen, there is plenty of fuel. Assuming that 10% of the sun's mass is in its core (the conditions for the nuclear reactions exist only in the core), we can estimate the total energy available to the sun. Dividing by the total energy by the sun's luminosity, we find that the sun can shine by this mechanism for about ten billion years. This is about twice the current estimate that evolutionists have for the age of the solar system, and hence the sun. Therefore, unlike with the Kelvin-Helmholtz mechanism, there is no conflict between the supposed age of the sun and the maximum age of the sun.

However, some creationists do not want to let this go. At a meeting of the American Astronomical Society in 1980, two astronomers presented evidence that the sun might in fact be shrinking. For two centuries, astronomers at the Greenwich Observatory in England had determined local noon nearly every clear day. They did this with a transit telescope, a telescope that can move only along the celestial meridian, a line passing north-south through the zenith. Observers would watch the image of the sun as it passed crosshairs on the telescope. They would measure the time when the sun's right limb passed the vertical crosshair and the time when the sun's left limb would pass the crosshair. The average of those two times is local noon.[3] Hidden in the two centuries of data was possible measurement of the apparent size of the sun, which can be found by subtracting the time measurements each day. The two astronomers presented their analysis of this data, which showed a slight decrease in the sun's apparent diameter over two centuries.

If the sun is powered by nuclear fusion, we would not expect its size to change over just two centuries. However, if the sun were powered by the Kelvin-Helmholtz mechanism, the sun must slowly shrink, so it might be possible to detect a slight decrease in size on this timescale. This startling result understandably gained much attention among biblical creationists. If the sun could be shown to be shrinking, then that would seem to confirm the Kelvin-Helmholtz mechanism, and eliminate nuclear fusion, as the source of the sun's energy. This would amount to a sort of silver bullet against evolution, because the world could not be billions of years old.

Oh, if it were only true! Other astronomers quickly responded with evidence that the sun was not shrinking. For instance, some pored over historical eclipse data. Some solar eclipses of the past were expected to be just barely total, right on the verge of being annular.[4] However, if the sun were slightly larger in the past, then these eclipses would have been annular rather than total. Historical records show that these eclipses were total. If the sun is not shrinking, how do we explain the data presented in 1980 that started this discussion? The data were very "noisy,"[5] and they showed only a slight decrease in solar diameter.

---

[3] Technically, this is apparent local noon. Due to the tilt of the earth's axis and the earth's elliptical orbit around the sun, the sun usually runs a little fast or a little slow. Astronomers define the mean sun as a fictitious average sun that moves at a uniform rate. The difference in time between the real and mean sun is the equation of time. The application to the equation of time permits knowing local mean noon from measurement of apparent local noon. Determining local mean noon is very important in timekeeping and navigation.

[4] See Chapter 4 for a discussion of eclipses.

[5] There are variations, or errors, in measurements that we make. Hopefully, the errors are small compared to the numerical value of the thing that we trying to find, which in this case is the sun's diameter. If so, then the errors of measurement are of little consequence. However, sometimes, as in this case, the errors are large compared to the quantity in question, making it difficult to know the precise value of what we are trying to find. This is analogous to trying to hear a faint sound amidst much noise, so when the errors are high, we say that the data are noisy.

Furthermore, virtually all the decrease in the sun's diameter occurred in two abrupt episodes. The epochs of these abrupt changes corresponded to two times that the telescope used to collect the data was refurbished. Even though the glass in the lenses of the telescope may have been the same, the dismantling and reassembly of the telescope at these two epochs probably slightly altered something within the telescope to cause slight changes in the sun's apparent size when there was no real change in solar diameter. The 1980 presentation was at best a preliminary study. At the time the two authors stated their intention to further develop and present their results, but they never did. Apparently, they quickly were convinced that spurious data had misled them. Unfortunately, while there is no clear evidence that the sun is shrinking, a few creationists persist in believing that it does shrink.

Related to the shrinking sun is the *solar neutrino problem*. Neutrinos are fascinating particles. As the name suggests, neutrinos have no charge. They have almost no mass, and they travel nearly at the speed of light. First hypothesized in 1930, their existence was not confirmed until 1956. The reason for the delay in the detection of neutrinos is that they hardly ever interact with other matter. Every second, a huge number of neutrinos pass through your body, as if you were not there. But on average, your body will stop one neutrino during your lifetime. Consequently, constructing an effective neutrino detector is not easy. Depending upon a neutrino's energy, it is slightly more likely to interact with certain isotopes. Therefore, physicists build neutrino detectors by amassing large amounts of isotopes that are good targets, effectively stacking the deck in favor of detection. The required isotopes often are exotic, making neutrino detectors expensive to build (this is in addition to the sophisticated scientific equipment required).

Many nuclear reactions, including those that power the sun, produce neutrinos. Since neutrinos pass through most matter as if it were not there, detection of neutrinos produced by reactions in the solar core offer the opportunity to probe directly the sun's core. We cannot do this optically, because light created in the solar core is absorbed and reemitted many times before it reaches the sun's surface. Consequently, it takes many years for light to reach the sun's surface, whereupon it takes only eight minutes for that light to travel from the sun's surface to the earth. But solar neutrinos, unencumbered by their passage through the sun's interior, arrive at the earth only eight minutes after their creation in the solar core.

The first solar neutrino detector was built in the mid-1960s. By the late 1960s, it was clear that something was wrong—the experiment was measuring only about a third of those predicted by the standard solar model. This difficulty came to be called the solar neutrino problem. The first solar neutrino detector measured neutrinos coming from a side reaction that accounted for only about 1% of the sun's energy. Could it be that for some reason that one reaction was shut down while the main reactions continued unabated? That did not seem likely, but by the 1980s other experiments went online to detect neutrinos coming from other reactions. They found similar results, indicating a shortfall from the predicted flux of solar neutrinos. Finally, experiments designed to detect neutrinos from the main reactions powering the sun began operation, and they too recorded a deficit by about two-thirds. Unlike the earlier solar neutrino experiments that could record detections but not the directions that neutrinos were moving, many of the later experiments measured direction of motion. They showed that most of the detected neutrinos indeed were coming from the sun. However, the measurements all showed that we were receiving only about one-third of the neutrinos expected from the sun.

Again, many biblical creationists seized upon this as evidence that the sun was shrinking. They reasoned that if the sun is deriving only one-third of its energy from nuclear sources, then the other two-thirds must be provided by some other means, most likely the Kelvin-Helmholtz mechanism. They thought perhaps God originally created the sun with no nuclear reactions going on in its core, but that those reactions had gradually begun to settle in over a few thousand years. That way, the sun still could be shrinking; so the two ideas, shrinking sun and the solar neutrino problem, were linked. And, more importantly, the sun could not be billions of years old, so evolution cannot be true.

Such a silver bullet argument would be nice, but, alas, there is no silver bullet. There had been two theories of the nature of neutrinos. The theory preferred by most particle physicists was that neutrinos have no mass, and hence travel at the speed of light. In this model, neutrinos are stable. The alternate theory was that neutrinos have very tiny mass and travel at nearly the speed of light. In this model, neutrinos are not stable, but instead oscillate between the three types of neutrinos, the electron type, the tau type, and the muon type. The sun produces only the electron type of neutrino, and the experiments designed to detect them measure only the electron type. But if neutrinos oscillate between the three types of neutrinos, then only about a third of them would remain the electron type by the time that they reach the earth eight minutes after their creation. This would explain the solar neutrino problem very well, so astronomers favored this solution to the problem, but, as previously mentioned, particle physicists disliked this solution. In 2001, experimental proof that neutrinos oscillate came, thus solving the solar neutrino problem. Therefore, the solar neutrino problem and the supposed shrinking sun are *not* a problem for a sun that is billions of years old.

Some creationists have attempted to undermine the conventional understanding of the sun another way. *Helioseismology* is the study of acoustic waves in the sun, that is, natural modes of vibration that exist in the sun's interior. These waves extend deep within the sun, so they offer the opportunity to probe the sun's interior, much as geophysicists can use seismic waves on the earth to probe the earth's interior. Some of the early helioseismology work four decades ago suggested a problem with our understanding of increasing density and temperature with depth inside the sun. This might have indicated that the sun's core is not nearly as hot as thought, and hence it might not sustain nuclear reactions. However, the resolution of the solar neutrino problem and the direct detection of many neutrinos coming from the sun argue strongly against this conclusion. Furthermore, more recent helioseismology studies confirm the standard model of the sun's interior.

Why did God make a nuclear-powered sun that can shine for billions of years, when the creation is not nearly that old? Part of the answer may lie in the sun's stability. The sun is remarkably constant in its size and surface temperature, and hence its luminosity. If gravitational contraction powered the sun, the sun likely would be less stable. This would cause some fluctuation in the sun's luminosity, which would be problematic for life on earth, as the earth's climate would change in response. However, the long timescale involved in nuclear reactions provides for a sun that is very stable. Even if the sun *could* last for billions of years, it does not mean that it has.

The Faint Young Sun Paradox

While direct confirmation of the sun's nuclear power source via the detection of neutrinos may allow for a sun that is billions of years old, that very source of energy is a problem for a sun and earth that are billions of years old. The fusion of hydrogen into helium in the sun's core gradually will transform the composition in the core. Over the ten-billion-year maximum lifetime of the sun, the core's composition ought to shift from about 75% hydrogen to nearly all helium. This change will slowly alter the mean molecular weight of the core, which in turn will cause a subtle increase in temperature in the core. The nuclear reactions are extremely sensitive to temperature, so as the temperature of the core increases, the nuclear reaction rate will increase. This will show up as an increase in solar luminosity. How much will the sun's luminosity change? Calculations reveal that in the supposed 4.5 billion years since the sun formed, the sun ought to have brightened by 40%. Since life supposedly arose on the earth 3.5 billion years ago, the sun would have brightened by 25%.

How much would a 25% increase in solar luminosity change the average temperature of the earth? It is relatively easy to show that the earth's temperature would have increased by 17°C. The average temperature of the earth now is 15°C, so the average temperature of the earth when life supposedly arose would have been -2°C. Since this average temperature is below freezing, the earth would have been largely ice-bound. Given the high reflectivity of snow and ice, if the earth ever became so entirely iced over, it is unlikely that the earth could have recovered. Hence, it probably never was this cold. Indeed, most paleontologists and most evolutionary biologists think that the earth's average temperature has fluctuated slightly over the ages, but that it has maintained something close to the current average temperature. However, the well-established physics of the sun's energy source precludes this. This problem is called the faint young sun paradox.

This problem has been known for a half century, so many solutions to the faint young sun paradox have been proposed. Some proposals suggest that the early earth was heated internally to offset the early sun's lower luminosity. One possibility is more radioactive decay early in the earth's history. Another possibility is that since the moon was much closer to the earth in the past, there was more tidal heating in the early earth. Other proposals revolve around other changes in the young sun that might have compensated for decreased luminosity when the sun was young. For instance, astronomers think that stars like the sun, but much younger than the sun, have much stronger winds. So, billions of years ago, the solar wind might have been much stronger than it is today. This could have played out at least two ways. If the early solar wind carried off 5–10% of the sun's mass, then the early sun would have been more massive than it is today. For most stars, there is a strong correlation between mass and brightness, so when the sun was more massive, it would have been brighter, perhaps compensating for the otherwise decreased luminosity when the sun was much younger. Within this scenario, as the sun matured, its lower mass resulted in lower luminosity, while the changing composition in the sun's core produced an offsetting increase in luminosity. The second solution relies upon the cooling effect that cosmic rays have on the earth. If the solar wind was much stronger in the past, it would have prevented cosmic rays from entering the portion of the solar system where the earth resides, thus preventing cooling from cosmic rays. As the solar wind diminished, more cosmic rays reached the earth, resulting in more cooling, again compensating for the increasing solar luminosity.

However, most proposed solutions to the faint young sun paradox rely upon gradual changes in the earth's atmosphere. If the earth's atmosphere had more greenhouse gases 3.5 billion years ago than it does today, then global warming would have heated the planet to maintain an average temperature similar to today, even though the sun was much fainter then. According to this scenario, the earth's atmosphere evolved so that a decreasing greenhouse effect compensated for the sun's increasing luminosity. Many people suggest that the appearance of living things played a major role in this: as life appeared, flourished, and evolved on earth, it altered the earth's atmosphere. Thus, the term *biosphere* has more meaning behind it than normally intended. The most extreme manifestation of this solution probably is the *Gaia hypothesis* proposed in the 1970s. Named for the primordial Greek goddess of the earth, the Gaia hypothesis proposes that the biosphere is a synergistic system that self-regulates many processes to maintain stability. The Gaia hypothesis easily can shift into pantheism, and indeed some of its adherents view this solution in spiritual terms.

There are several salient observations about these proposed solutions for the faint young sun paradox. First, there is no evidence that any of them operated in the past. Instead, these solutions involve plausibility arguments supported by various bits of data. However, plausibility does not necessarily correlate with reality. Second, each of these solutions ask that we believe that two unrelated processes operating over billions of years without any feedback mechanism worked in tandem to cancel out one another. That would have been remarkable, which probably explains the theistic or spiritual connection that some people see, as with the Gaia hypothesis. Third, the multitude of proposed solutions amounts to a tacit admission that none of them effectively solve the faint young sun paradox. Over the years, I have watched as each new solution is published, is hailed as *the solution* to the faint young sun paradox, and then is followed by the same cycle for the next new solution. Left unsaid in most cases is the need for a new solution—if the previous solution worked so well, why propose a new one? The faint young sun paradox exists only in a world that is billions of years old. But if the world is only thousands of years old, there is no paradox to resolve. If the sun is young, then there has been no evolution in the composition of the solar core. *Problem solved.*

CHAPTER 7

He Calls Them All by Name:
Stars and Stellar Distances

I have already mentioned that the stars are much farther away than the sun, but how far away are they? The nearest star is 26 trillion mi, or 40 million km, away. Those huge numbers are difficult to comprehend, so perhaps we ought to use a larger unit to express distance, such as the astronomical unit (AU). The nearest star is 275,000 AU away, and most stars are over a million AU away. While those numbers are smaller, they still are daunting, so when dealing with the general public, astronomers typically express stellar distances using the light year, the distance that light travels in a year. Multiplying the speed of light (186,282 mi per second [299,792 km per second]) by the number of seconds in the year, we find that the light year is about six trillion miles. The nearest star is 4.3 light years away. Some of the stars we see at night are much farther—hundreds or, in a few cases, thousands of light years away. Contrast that to the sun, which is only eight light minutes away.

How Do Astronomers Measure Stellar Distances?

But how do astronomers *know* these distances? The answer to that question takes some explanation. In Chapter 1, we saw that ancient astronomers realized that if the earth orbited the sun, they ought to observe parallax. Since they did not see any parallax, most ancient astronomers rejected the heliocentric theory. The few ancients who did believe the heliocentric theory did so despite the best evidence then available, arguing that the stars were so distant that stellar parallax would be too small to measure. They turned out to be correct.

The largest parallax is 0.76 arc seconds,[1] the apparent size of a dime when viewed from three miles. Obviously, the naked eye cannot discern such a small angle, and it was not until the 1830s, long after the invention of the telescope, that astronomers first measured parallax. While still difficult today, parallax measurements have come a long way since.

Parallax is relevant to stellar distance because the amount of a star's parallax is related to its distance (Figure 7.1). The circle in the diagram represents the orbit of the earth. The line from the sun to the star represents the distance to the star. The distance to the star makes a right angle with the line connecting the sun to the earth. These two lines form the legs of a right triangle. Note that the shorter leg, the line between the earth and sun, is 1 AU long. This will be true for the triangles formed by every star, even though the other legs of the triangles will be different, because stars are at different distances. Therefore, the earth's orbital radius makes a convenient base for measuring stellar distances. There is a small angle, π, formed by the distance to the star and the hypotenuse of the triangle, and we usually call this angle the parallax.[2] As the star's distance

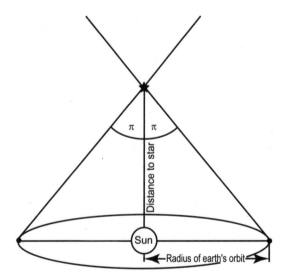

Figure 7.1. Illustration of stellar parallax.

---

[1] A degree of arc is divided into 60 minutes, and each minute is divided into 60 seconds. Therefore, an arcsecond is $1/_{3600}$ of a degree.

[2] Notice that π here does not refer to the irrational number representing a ratio of a circle's circumference to its diameter. Rather, it is an *angle*, using the conventional practice of naming angles for Greek letters. Astronomers selected the letter π, because it is equivalent to the Latin letter p, which is the first letter of the word *parallax*. It is important to realize that π is a variable, much as x often is the variable in algebra.

increases, the angle π decreases. That is to say, the distance and parallax are inversely proportional. Being a triangle, one normally would express the relationship between distance and parallax with the tangent of the angle and the distance, but because the distances involved are so large and the angles so small, this is unwieldy. It is far more convenient to take advantage of the inverse proportionality to relate the distance, d, to the parallax:

$$d = 1/\pi.$$

This requires using compatible units, so astronomers define the following: parallax measurements are in seconds of arc, while the distance is in a new unit called the *parsec* (abbreviated pc). The word parsec is a portmanteau of the words *parallax* and *second*, because, as you can see from the equation, a star that is one parsec away will have a parallax of one arc second. One parsec is approximately 3.26 light years, so the closest star is 1.3 pc away. While astronomers may use light years in dealing with the public, in the technical astronomical literature the pc is preferred. Since the closest star has a parallax of 0.76 arcsecond, no star has a parallax as large as even one second of arc. Parallax measurements typically are reported in milliarc seconds (mas), the milliarcsecond being one-thousandth of an arcsecond.[3]

Parallax measurements are made six months apart, when the earth is on opposite sides of its orbit around the sun. Stars within any given part of the sky are at the proper location for parallax measurements twice per year. The prime location for parallax measurements continually changes throughout the year depending on the star, so astronomers who specialize in parallax studies always have program stars to measure. This sort of work is very tedious. Originally, astronomers made visual measurements with an instrument called a filar micrometer attached to a telescope. Once photography became available, it became the standard method. However, measuring positions of stars on pairs of photographs taken six months apart was very time consuming. These traditional techniques permitted accurate measurement of stellar distances (within 20%) out to about 20 pc (65 light years). There are about 300 stars within this distance.

---

[3] An arcsecond is one-sixtieth of an arcminute, and an arcminute is one-sixtieth of a degree. Therefore, an arcsecond is $1/_{3600}$ of a degree.

Modern technology has greatly improved upon the classical techniques. In 1989, the European Space Agency (ESA) launched the Hipparcos[4] satellite on a four-year mission to measure parallax to unprecedented accuracy. It was the combination of the new techniques of observations by this spacecraft and being above the blurring effects of the earth's atmosphere that made this possible. The *Hipparcos Catalogue*, with high precision distance measurements for nearly 120,000 stars, was published in 1997, along with the *Tycho Catalogue*, which had less precise distance measurements of more than a million stars. The *Tycho-2 Catalogue*, with distance measurements of 2.5 million stars, was published in 2013. The *Hipparcos Catalogue* extended 20% precision of stellar parallaxes an order of magnitude (to about 600 light years) over what was possible with classical techniques from the ground.

As impressive as the results of Hipparcos have been, it is soon to be replaced by the data from the Gaia spacecraft. Launched by ESA in 2013, over its five-year mission Gaia is expected to measure accurately the positions of about one billion astronomical objects. With the position of each of these objects measured about 70 times, the parallax measurements ought to improve upon the Hipparcos distance determinations by another order of magnitude. That is, we will know the distances of most stars within 6000 light years to an accuracy of 20%. Of course, stars closer than 6000 light years will have more accurately determined distances. The data will be released in stages, with the first release having occurred before the publication of this book.

Trigonometric parallax is the only direct method of measuring the distances of stars. While complicated in actual practice, the basic theory behind it as described above is readily understandable. Furthermore, it is identical to surveying techniques that have been used on earth for a very long time. Therefore, we have much confidence in the results of trigonometric parallax measurements. However, trigonometric parallax has always been subject to an upper limit on distance, and astronomers have desired to know the distances of stars beyond this limit, so they have developed alternate indirect methods. This is not the place to discuss all, or even most, of these methods, but I will describe one of the more important ones here, the use of *Cepheid variables*.

---

[4] Hipparcos is an acronym for **High precision parallax collecting satellite**. The acronym was selected as a variation of the spelling of the second-century BC. Greek astronomer Hipparchus. The mission's name was intended to honor Hipparchus, who was the founder of trigonometry and did the earliest work on positional astronomy.

Variable stars are stars that change brightness rather than remaining constant. Astronomers recognize many different types of variable stars. One type, Cepheid variables, are named for the prototype, the first one studied in detail, δ Cephei (Figure 7.2). Cepheids (what astronomers often call Cepheid variables) are very large stars—they are among stars that astronomers call giants and supergiants. They have surface temperatures close to that of the sun. Cepheids vary in brightness, because they pulsate, alternately expanding and contracting. As with any gas, as Cepheids expand and contract, their temperatures change slightly. The brightness of a star depends upon both the star's size and temperature. To illustrate, let L be a star's luminosity (brightness), R be its radius, and let T be its surface temperature, expressed in Kelvin (K). Then the star's luminosity is,

$$L = \sigma\, 4\pi\, R^2 T^4,$$

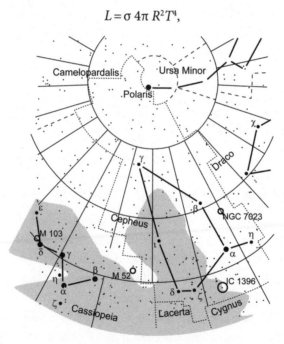

Figure 7.2. Star chart highlighting δ Cephei. (Wikimedia Commons).

where σ is the Stefan-Boltzmann constant. As you can see, changes in a star's radius and temperature will result in changes in the star's brightness. A *light curve* is a plot of a variable star's brightness over a complete cycle. A representative Cepheid light curve is illustrated in Figure 7.3. Cepheids have distinct light curves, with a steep rise from minimum brightness to maximum brightness, followed by a much more gradual decline back to minimum brightness. The period is the length of time required for a variable star to go through one complete cycle of light variation. With a Cepheid, this is most practical to express as going from one time of maximum light to the next. The periods of Cepheid variables are very stable, so observations over several years can be used to determine the period of any Cepheid precisely. The average brightness is defined as the average of the maximum and minimum brightness.

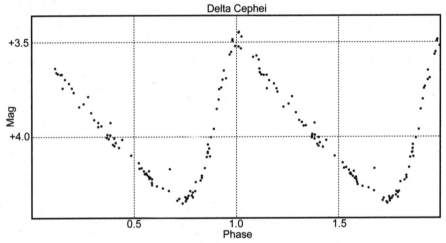

Figure 7.3. Representative Cepheid light curve. (Wikimedia Commons).

A century ago, the astronomer Henrietta Leavitt, while studying Cepheids in the Large Magellanic Cloud, a satellite galaxy to our Milky Way Galaxy, discovered that there was a relationship between the average brightness of Cepheids and their periods, with the longer-period Cepheids being brighter. For a while this was called Leavitt's law, but now it usually is called the Cepheid period-luminosity relationship. Leavitt immediately recognized how useful this was, because by measuring the period of a Cepheid, she could determine its intrinsic brightness. The measured brightness depends upon the both the intrinsic brightness and the distance, so knowing the measured and intrinsic brightness allows us to compute the distance. All astronomers needed was the distance to at least one Cepheid to calibrate this method. Unfortunately, there were no Cepheids close enough to measure their distances directly using trigonometric parallax. For many years, astronomers used another indirect method to calibrate the period-luminosity relation of Cepheids. This method, called statistical parallax, relied upon applying statistical analysis to brightness and motion measurements of a large sample of Cepheid variables. That technique was refined over the next 75 years. There are several Cepheids within 600 light years, so the Hipparcos mission was the first time that the Cepheid period-luminosity relation was calibrated by direct measurement of distances. As it turned out, the previous indirect calibration was off by less than 5%, giving confidence to both the Cepheid period-luminosity relation and the method of statistical parallax.

How Do Astronomers Measure Stellar Brightness?

Measuring the brightness of astronomical objects would seem to be straightforward, but it is not. Astronomers use a system of magnitudes to measure stellar brightness. This practice goes back two millennia, to the Greek astronomer Hipparchus. Hipparchus grouped the 20 brightest stars in the sky into the first magnitude. The second tier of fainter stars he grouped as the second magnitude, and so forth up to the sixth magnitude, the faintest stars visible to the naked eye on a clear, dark night. Notice that this system is backwards from the way that most measurements work, with the smallest number of measure corresponding to the greatest quantity being measured. This is consistent with the ancient Greek idea of brighter stars being more important. We do the same thing today with how we classify hotels and such as first class, second class, etc. Furthermore, this system is not linear, where a doubling in the measured quantity results in doubling the numerical expression of the measurement. Rather, this system is logarithmic, because it is based upon what the eye perceives, and human sense perceptions, including vision, are approximately logarithmic responses. In 1855, the English astronomer Norman Pogson noted this fact and observed that first magnitude stars are about 100 times brighter than sixth magnitude stars, a difference of five magnitudes. Therefore, Pogson proposed that we define a ratio in brightness of 100 to correspond exactly to a difference of five magnitudes. Mathematically, this is expressed as,

$$\Delta m = -2.5 \log_{10} (I_2/I_1),$$

where Δm refers to a difference in magnitude, $m_2 - m_1$, and I_2 and I_1 refer to the brightness of two different stars. The negative sign is required by the fact that the magnitude system is backwards, with brighter stars having a lower number than fainter stars. This equation defines the Pogson scale of magnitudes.

The invention of the telescope revealed many stars too faint to be seen with the naked eye; hence astronomers have long recognized stars that have higher magnitude numbers than six. And we have extended the system to brighter objects, such as the sun and moon, forcing the use of numbers less than zero. For instance, the full moon is −12.7 magnitude, while the sun is −26.7. There is even a range in brightness among the first magnitude stars defined by Hipparchus that requires use of negative numbers. Sirius, the brightest appearing star, is −1.46. The brightest planet is Venus, which can shine at −4.5, while Jupiter, the second brightest planet, can be as bright as −2.9. Jupiter and Venus usually are fainter than their maximum magnitudes. Their brightness changes for several reasons, one being that their distances from the earth change as both we and they orbit the sun.

This underscores an important point: the magnitudes that we measure for stars are based upon the amount of light that we receive from them on earth. We define this as the *apparent magnitude*. However, stars are at various distances from us, so the apparent magnitude does not indicate a star's intrinsic brightness. Astronomers use absolute magnitude to express a star's intrinsic brightness. Absolute magnitude is defined as the apparent magnitude that a star would have if it were a standard distance of 10 pc away. Let m be the apparent magnitude, M be the absolute magnitude, and d be the distance in pc. The three quantities m, M, and d are related via

$$m - M = 5 \log_{10} (d) - 5.$$

We call the difference between apparent and absolute magnitude, $m - M$, the distance modulus. This equation is useful in finding the absolute magnitude if we measure the apparent magnitude and we know the distance of the star, say by trigonometric parallax. What if we measure the apparent magnitude and know the absolute magnitude, such as by use of the Cepheid period-luminosity relation? We can solve the above equation to yield,

$$d = 10^{(m-M+5)/5}.$$

Notice that each difference of five in distance modulus results in a factor of ten in distance. This is not happenstance. Recall that the modern magnitude system is defined as a difference in five magnitudes corresponding to a factor of 100 in brightness. The amount of light is proportional to the inverse square of the distance, so a change in brightness of 100 corresponds to a change in distance of 10.

The output of stars is not uniform at all wavelengths, but rather is a complicated function of wavelength. Therefore, what magnitude we measure depends upon at what wavelength we make the measurement (Figure 7.4). This was not a problem throughout most of history, when magnitude measurements were made with human eyes. However, by the latter part of the nineteenth century, photography became the preferred method of measuring stellar magnitudes. Magnitudes measured photographically were more accurate than those made by the eye, because one can measure the size and density of the spot recorded on the photographic negative rather than estimating with one's eye. However, the eye's sensitivity peaks in the yellow-green part of the spectrum, while the original black and white photographic emulsions were very blue-sensitive. This came into stark contrast with the stars of the constellation Orion. Most stars in Orion are very blue. The one notable exception is Betelgeuse, the brightest star in Orion, at least as determined by what the eye sees. But on most photographs, Betelgeuse appeared much fainter than the other stars,

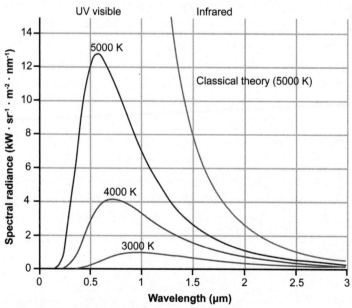

Figure 7.4. Blackbody graph. (Wikimedia. Public Domain).

because it produces little light in the blue part of the spectrum. Eventually, yellow, and even red photographic emulsions were developed. By the middle of the twentieth century, photoelectric detectors began to replace photography for precise magnitude measurements, because photoelectrically measured magnitudes were more accurate. By the 1980s, charge coupled devices, or CCDs, began to replace both photographic and photoelectric measurements of magnitudes. CCDs are computer chips with a large array of very tiny photosensitive spots, or pixels. The advantage of CCDs is that they are nearly 100% efficient, recording virtually all light that falls on them. Compare that to 1–2% efficiency of photographic emulsions and 15% efficiency of photoelectric cells.

Astronomers define different apparent magnitudes, each with their own band passes[5] as a portion of the optical spectrum. Each defined magnitude is a combination of the wavelength sensitivity of the detector used and the color of a filter placed in front of the detector. For instance, the V (standing for *visual*) magnitude closely matches what the human eye detects, but the B filter mimics what the original standard blue-sensitive black and white photography measured. These defined magnitudes permit easy comparison between currently measured magnitudes and ones from the past using entirely different equipment. Furthermore, the difference between the B and V magnitudes, B-V, allows a very quick determination of a star's temperature.[6] Finally, comparison of observations and theory often requires expressing a magnitude including energy radiated at all wavelengths. Astronomers call this the bolometric magnitude, usually expressed as an absolute bolometric magnitude, M_{Bol}. It is not possible to measure the light from a star at all wavelengths, but astronomers can measure the magnitude at various portions of the spectrum and then apply a bolometric correction accurately to estimate the bolometric magnitude. Astronomers construct models, or theories, to describe the physical structure of stars. One of the outputs of a stellar model is luminosity, the entire energy radiated by a star over all the wavelengths. We can convert this luminosity to absolute bolometric magnitude to compare with observations.

---

[5] A band pass is the portion of the spectrum that is measured.

[6] This is because the B and V bands sample different parts of the spectrum. Stars of different temperature have different fluxes in different parts of the spectrum.

Binary Stars: The Key to Measuring the Masses and Sizes of Stars

Mass is the measure of how much matter an object has. On earth, we usually measure an object's mass by measuring its weight, the force of gravity that the earth has upon the object. From Newton's law of gravity, we know that the force of gravity on an object is proportional to the object's mass, so it is easy to convert the force of gravity to mass. In fact, some scales and balances are calibrated in mass rather than weight,[7] so that no conversion is necessary. But how does one ask a star to step on a scale so that we can measure its mass? Obviously, this is not possible, but to measure a star's mass, all that we need is a measure of the gravity it exerts. Binary stars wonderfully provide this opportunity.

What is a binary star? A binary star is a system of two stars orbiting one another via their mutual gravity. This is similar to how planets orbit the sun. As the sun's gravity pulls on the planets, by Newton's third law of motion, the planets pull back with equal force. However, because the sun has so much more mass than the planets have, the planets move quite a bit while the sun moves very little. The situation is very different with binary stars. The masses of the two stars are comparable, so both stars move noticeably. We can use this information to our advantage. If we carefully measure the motion of the two stars involved, we can use physics to measure the masses of the two stars. Most stars are members of binary star systems, so we are not at a loss for stars for which we can measure their masses. Why are there so many binary stars? No one knows. One possibility is the implication of design. God may have made so many binary stars so that man could measure stellar masses, a fundamental quantity vital in probing the workings of stars.

We can divide binary stars into two broad classes, defined by how we observe them. If the orbits of the two stars are very large and the system is not too far away, we can see the individual stars through a telescope. This is a *visual binary*. Over time, we can see the positions of the two stars in a visual binary

---

[7] In everyday use, many people confuse mass and weight. The difference between weight and mass can be subtle yet profound. Newton's second law, $F = ma$, relates force to mass and acceleration. Since weight, W, is a force, we can substitute it, as well as the acceleration of gravity, g, into Newton's second law: $W = mg$. Notice that once one knows the weight, division by the acceleration of gravity yields the mass. The metric unit of force is the Newton, and the unit of mass is the kilogram. In the English system, the pound is the unit of force and the slug is the unit of mass. Since the kilogram is a unit of mass, and the pound is a unit of force, the kilogram and pound do not measure the same thing. However, if all measurements are made with the same gravitational acceleration, we have the luxury of sloppily confusing the two.

change as they orbit one another. If we observe them as they travel through a complete orbit, or at least a major part of their orbit, we can determine the mass of either star. Unfortunately, to see the individual stars, the stars in the binary must have very wide separation. But very wide visual binaries have very large orbits and hence very long orbital periods, so it may take decades of observations to have enough data to measure the masses of the stars. And if the orbits are so large as to take centuries or millennia to orbit, it is difficult to determine the stars' masses accurately over even decades of observation.

If the stars in a binary system orbit much more closely, the orbital period is much shorter, often a matter of weeks, days, or even a fraction of a day. However, when orbiting so closely, we cannot see the individual stars, even when observed through the largest telescopes. Instead, their light blends together into what appears to be a single star. The binary nature is betrayed by the Doppler motion of the two stars as they orbit one another. The Doppler effect is a shift in the spectrum of the light given off by an object as it moves.[8] Since the two members of a binary star system alternately move toward and away from us as they orbit one another, there are periodic Doppler shifts in the spectrum of the system. This is called a spectroscopic binary. Observations of the spectrum over an entire orbital period produce a radial velocity curve. Analysis of the radial velocity curve yields the mass function, involving the mass and the cube of the orbital inclination.[9] If the stars have similar brightness, we can determine the mass function of either star. However, if one star is more than three times brighter than the other star, then we see the spectrum of only one star, and we can determine the combined mass function, which involves the sum of the masses and the cube of the orbital inclination.

If we know the orbital inclination, we can compute the masses of the stars in a spectroscopic binary. However, in most cases the inclination is not known. Statistical analysis of many similar systems can result in a statistical average for the inclination, which in turns allows computation of a statistical average mass

---

[8] You probably have experienced the Doppler effect with sound. If you or a source of sound is moving, the pitch that you hear changes. This is because relative motion toward effectively shortens wavelengths, and relative motion away stretches the wavelengths. The same thing happens with light.

[9] Orbital inclination is defined as the angle that the orbital plane makes with the plane of the sky, which is the plane perpendicular to our line of sight to the binary. Therefore, if we look directly down onto the orbital plane, the inclination is 0°, and if we view the orbit edge on, the inclination is 90°.

for similar stars. These results agree with masses determined from systems containing similar stars where the inclination is known. How can orbital inclination be determined? A subset of spectroscopic binaries are *eclipsing binaries.* An eclipsing binary is a binary star system where the orbital inclination is near 90°, so that two stars undergo mutual eclipse each orbit. Measurement of brightness over an entire orbital cycle (or phased when observations over several cycles are combined) produces a light curve. Analysis of the light curve accurately indicates the orbital inclination, from which the combined mass or the individual masses of the stars are revealed.

Eclipsing binary stars have the added advantage of providing the only direct method of measuring the sizes of stars. The duration of the eclipses in an eclipsing binary system depends upon the sizes of the two stars. Spectroscopic observations reveal the speeds that the stars are orbiting one another. Therefore, if we have spectroscopic data, then we can multiply the duration of the eclipses by the orbital speed to get the sizes of the stars.

In addition to this direct method of measuring stellar sizes, there are several indirect methods. They are considered indirect because these methods require use of additional information, primarily the distance from earth, that has nothing to do with the actual sizes of stars. For instance, the moon occasionally passes in front of a bright star, an event that we call a *lunar occultation.* The star's disappearance and subsequent reappearance behind the moon's limb is very rapid, but it is not instantaneous. The length of time required for the star to disappear or reappear depends upon several factors, including the angular diameter of the star.[10] Astronomers can account for the other factors, so measurement of the time required for the star to disappear or reappear allows calculation of stellar angular diameters. This method of angular diameter measurement works only for brighter stars that lie in the relatively small swath of sky that the moon appears to pass through as it orbits the earth, so it is of limited use.

Another method of angular diameter measurement uses an interferometer, a device that measures the interference patterns produced when the light of a bright star from two telescopes is combined. The interference pattern depends upon several factors, including angular diameter; but since the other factors are

---

[10] Angular diameter is how large an angle that the disk of an object appears to subtend. For stars, this normally is expressed in milliarcseconds. In contrast, the moon itself subtends an angle of about ½°, or 1800 arcseconds, or 1.8 million mas.

known, the angular diameter easily follows. A star's angular diameter depends upon the star's diameter and distance, so if we know the distance, we easily can compute the diameter.

One other indirect method of measuring stellar size does not depend upon knowing the angular diameter. Instead, it relies upon knowing a star's temperature and luminosity. Stellar temperature can be measured several ways, such as knowing a star's B-V color. As previously mentioned, luminosity is an expression of a star's total energy output, which can be transformed from an absolute bolometric magnitude. Of course, absolute bolometric magnitude requires knowing the star's distance and the application of a bolometric correction to an apparent magnitude. Let L be a star's luminosity, T be the star's temperature, and R be the star's radius. Assuming that the star is a sphere, the luminosity is,

$$L = \sigma \, 4\pi \, R^2 \, T^4,$$

where σ is the Stefan-Boltzmann constant. Since we know the luminosity and temperature, we can solve for the size.

Spectral Types and the Hertzsprung-Russell Diagram

The spectra of most stars are crossed by dark lines at various wavelengths or colors. These lines are produced by atoms and ions of different elements in stellar photospheres absorbing light at specific wavelengths. Each element has a unique set of absorption lines, so the absorption lines can be used to identify which elements are present in stars, and in what quantities. In the 1880s, Harvard College observatory began a long-term systematic program of recording and classifying stellar spectra. At first, the physical mechanisms responsible for absorption line formation were not understood, and so a classification scheme emerged based solely upon which spectral lines were visible. This scheme started with the letter A and concluded with the letter Q.[11] The first publication was in 1890 and included classification of more than 10,000 stars. A more detailed study in 1897 dropped and combined many of the classifications. A southern sky extension in 1901 was the first to include a classification system close to the modern one. Over the next few decades, the Henry Draper Catalogue of stellar spectra was published, including more than 350,000 stars.

---

[11] Not, of course, to be confused with the Q Continuum.

The basic spectral types in the modern system are, in order of decreasing temperature, O, B, A, F, G, K, and M. This somewhat out-of-order system results from the early classification, based upon which spectral lines were visible and the later realization that which spectral lines are present in a star's spectrum depends on the temperature of the star's photosphere. Absorption of energy as electrons go from lower orbits to higher orbits around atomic nuclei causes spectral lines. This happens in the relatively cooler atmospheres of stars. Light coming from deeper layers within stars are simple blackbody curves lacking spectral lines. The very hottest stars have such high photospheric temperatures that all atoms are highly ionized, so that only a few atomic transitions are possible. Therefore, with temperatures exceeding 30,000 K, O-type stars have only a few absorption lines, due to a few elements, such as helium. With decreasing temperature, helium lines disappear and are replaced by strong hydrogen lines. Hydrogen lines peak at a temperature of 10,000 K, corresponding to A-type stars. With ever cooler temperatures, the strength of the hydrogen lines fade and are replaced by ionized metal lines. At even cooler temperatures, ionized metal lines are replaced with neutral metal lines, and then eventually they are joined by lines of molecules in the spectra of the coolest stars. Spectral types are further divided into subtypes, running 0 through 9. For instance, starting with A0, there are A1, A2,... A9, F0, F1,... G0, G1, and so forth. A0 is preceded by B9.

At first, astronomers thought that the presence or absence of absorption lines of elements in a star's spectrum merely indicated the presence or absence of those elements in the stars. However, by the early twentieth century, astronomers began to understand the physical conditions required for absorption lines to appear in stellar spectra. Since absorption lines form when electrons in a particular orbit around atomic nuclei absorb light and use the energy absorbed to ascend to a higher orbit, the absorption line can occur only if a sufficient number of electrons exist in a particular orbit within the star's photosphere. However, if a star's photosphere is too hot or too cool, there will not be enough electrons in the appropriate orbit around nuclei of a particular element to permit the absorption lines of that element. For instance, consider the hydrogen lines that exist in the spectra of many stars. There are three hydrogen spectral lines in the visible part of the spectrum. They are part of the Balmer series of spectral lines that continue into the ultraviolet, which the eye cannot see, but which we can detect by other means. The Balmer series absorption lines arise from transitions of electrons originally in the second orbit around hydrogen nuclei. The first line in the Balmer series, Hα, occurs when an electron goes from the second to third orbit. Hβ comes from a transition from the second to fourth orbit, and Hγ results from a transition from the second to fifth orbit. In very cool stars, nearly all of the electrons in hydrogen atoms are in the ground state, the first orbit, so there are not many electrons in the second orbit from which Balmer lines could be produced. In very hot stars, most electrons in hydrogen atoms are in very high orbits or are ionized. Therefore, there are not enough hydrogen atoms with electrons in the second orbit to produce significant Balmer lines. The greatest number of electrons in the second orbit of the hydrogen atoms exist at a temperature of about 10,000 K (spectral type A0), which is why the Balmer hydrogen lines are most pronounced in stars with that temperature. Stars progressively cooler or warmer than 10,000 K have correspondingly weaker Balmer lines.

Similar reasoning applies to absorption lines of other elements. We can use information about which spectral lines are present in a star's spectrum to measure the temperature of the star's photosphere. It is good to compare this independent measure of temperature with other measures, such as measures from color. Again, the absence of absorption lines of an element in the spectrum of a star does not necessarily mean that the element is absent in the star. Rather, the spectral lines may be absent if the temperature is too high or too low for the lines to form. If the temperature is conducive for the formation of spectral lines of an element but those lines are absent, then we may infer that the element is absent in the star. Furthermore, once the temperature of a star's photosphere is ascertained, we can use the strengths of the spectral lines present to gauge the relative abundance of various elements. That is, weak spectral lines where the temperature is conducive to formation of strong lines is almost certainly due to a low abundance of the element in question. In this manner, astronomers have determined the elemental abundance of the sun and many other stars. The composition of most stars is amazingly similar, with about three-quarters being hydrogen, about one-quarter being helium, and at most a few percent of everything else. A small percentage of stars have odd compositions. Astronomers have developed theories to explain both the similarity of composition for most stars and the few odd stars.

A little more than a century ago, the Danish astronomer Ejnar Hertzsprung and the American astronomer Henry Norris Russell produced plots of the absolute magnitudes of stars versus some measure of stellar temperature. One might expect such a plot to be a scatter diagram, but most stars fall along a roughly diagonal band. The Hertzsprung-Russell (H-R) diagram, as this plot came to be called, soon became the foundation for understanding stellar structure (Figure 7.5). On the accompanying diagram, you can see that intrinsic brightness increases upward. On an H-R diagram, intrinsic brightness can be expressed as luminosity or absolute magnitude. One can even use apparent magnitude, if only a group of stars of about the same distance, such as stars within a star cluster, are plotted. If different star clusters are plotted, then the vertical axes will not line up, but if we know the distances of each cluster, then the vertical scales can be shifted so that they match. The temperature scale of the horizontal axis can be expressed several different ways too. One way is to plot the temperatures of stars. Another way is to use B-V color. Since spectral type is related to temperature, spectral type can be used as well. On most plots, the quantity being displayed increases to the right. However, on the H-R

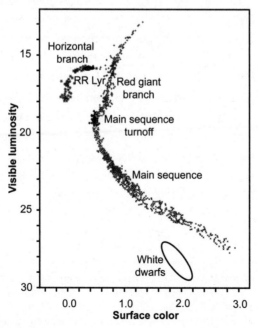

Figure 7.5. Basic H-R diagram.

diagram, the temperature scale increases toward the left. This unusual situation is a vestige of how the H-R diagram was first plotted, with relation to spectral type. At the time, no one knew that the spectral types in order represented a temperature scale. Unfortunately, the astronomers at the time selected a sequence of spectral types that went from hotter to cooler stars. Astronomers eventually figured this out, but it was too late to correct the sequence.

Notice on the plot that most stars fall along a roughly diagonal band from upper left to lower right. Astronomers call this the *main sequence*, and stars along the main sequence (MS) are called main sequence stars. This means that hotter MS stars are brighter than cooler MS stars. This makes sense, for it is consistent with the previous equation,

$$L = \sigma\, 4\pi\, R^2\, T^4.$$

From the equation, you can see that for a given size, the luminosity of stars increases with the fourth power of the stellar temperature. The hottest MS stars are about 10 times hotter than the coolest MS stars. Therefore, if the hottest and coolest MS stars were the same size, then we would expect that the hottest stars would be $10^4 = 10,000$ times brighter than the coolest MS stars. However, the hottest MS stars are millions of times brighter than the coolest MS stars, suggesting that the hottest MS stars are larger than the coolest MS stars. Indeed, the hottest MS stars are about 20 times larger than the faintest MS stars.

As previously discussed, the study of binary stars allows us to determine stellar masses. There is a very strong correlation between mass and the location of stars on the MS. Stars near the bottom of the MS have low mass, while stars near the top have high mass. In between, mass steadily increases from the bottom of the MS to the top. Since the MS has steadily increasing luminosity from bottom to top, this can be expressed as a mass-luminosity relation.[12] The mass-luminosity relation can be expressed simply as,

$$L = M^{3.5},$$

---

[12] We would expect this physically. Stars with more mass have more gravity. Since pressure in a stellar core must support the weight of overlying layers, greater gravity results in greater core pressure. Greater pressure generally corresponds with greater temperatures, so the cores of more massive stars ought to be hotter than less massive stars. The nuclear reaction rates that power stars are very temperature sensitive, so greater core temperature results in more nuclear fusion, and hence greater luminosity.

where L is the luminosity and M is the mass, with both expressed in solar units. That is, the luminosity and mass of the sun both are one; a star that has twice the luminosity would have luminosity of two, and a star with three times the mass of the sun would have mass of three. This equation is only an approximation of the mass-luminosity relation over the entire MS. This equation works well for the upper main sequence, but for stars similar to the sun, the exponent is closer to 4 than to 3.5. For low mass MS sequence stars, the exponent is close to 2.3. Alternately, since mass increases upward on the MS, one can define a mass-radius relation, though the mass-radius relation is not referenced nearly as often as the mass-luminosity relation is.

Neither the mass-luminosity nor mass-radius relations work for stars not on the MS. Consider stars to the upper right of the H-R diagram. These stars have cool surface temperatures, so by the equation that describes stellar luminosity, they ought to be faint stars, but they are not. The only explanation for this is that these stars are very large. Many of these stars are tens, hundreds, or even thousands of times larger than the sun. Astronomers call these stars giants. Now consider stars to the lower left of the MS. These stars are very hot, so from the equation that describes stellar luminosity, these stars ought to be very bright, but instead they are very dim. The only explanation for this is that these stars are very small. Because of their high temperature, these stars often appear white. Being so small and hot, astronomers call these stars white dwarfs. A white dwarf is 1–2% the size of the sun, or about one to two times the size of the earth. The density of a white dwarf is immense—a teaspoon of a white dwarf transported to earth would weigh a ton!

How does the sun compare to other stars? The sun's spectral type is G2, and it is a MS star. This places the sun near the middle of the H-R diagram. In some sense, this makes the sun average. There are stars much brighter and far larger than the sun, so the sun often is described as below average. However, most stars on the MS are lower than the sun's position. Therefore, in terms of mass, size, and luminosity, the sun is brighter than most stars. It is not clear why the sun is so often described as below average when it clearly is not.

Star Formation

As discussed in the Introduction to this book, most scientists today, including astronomers, are committed to naturalistic theories of origin. Very broadly, we can include naturalistic theories of origin under the heading of evolution. Therefore, the theory of how stars form and develop is called *stellar evolution*. Some people do not understand this, because they have always heard the word *evolution* used only in the context of *biological evolution*, and so they think that the word *evolution* applies only to living things. However, the word *evolution* is used in many contexts, including astronomy. Astronomers frequently and consistently refer to the formation and change in stars over time as stellar evolution. Perhaps a better term would be *stellar aging*, because this refers to changes in single stars. Comparisons to biological evolution are poor, because while stellar evolution refers to changes in individual stars, biological evolution involves changes in populations of organisms.

We saw in the previous chapter that the sun produces energy through nuclear fusion in its core, a process that could power the sun for about ten billion years. We think that most stars (certainly those on the MS) derive their energy by the same mechanism. Calculations reveal that stars on the lower part of the MS could exist for many tens of billions of years, but that stars near the upper end of the MS have very brief lifetimes, only a few million years. I shall discuss in Chapter 10 the big bang model, which proposes (as most astronomers believe) that the universe suddenly appeared 13.8 billion years ago. This age is greater than the maximum age of the sun and many other stars, so if the big bang model were true, the sun and these stars could not be primordial. Rather, they must have formed sometime after the big bang (in the case of the sun, this supposedly was 4.6 billion years ago, more than nine billion years after the big bang).

Where and how do stars form? There are many clouds of gas in space. Like stars, these clouds are made mostly of hydrogen. The mass of a gas cloud typically is thousands of times that of a single star, so most astronomers think that stars somehow form from these clouds. Being so massive, these clouds possess self-gravity, so most astronomers reason that this gravity might help the clouds to condense into stars. However, the gravity is very feeble, and the clouds possess gas pressure too. If a cloud were to contract slightly, the gas pressure would increase, causing the cloud to reexpand. That is, gas clouds that we observe are stable against collapse. This has been a long recognized problem. If a cloud could be compressed to a certain size, the *Jeans' length*, then gravity theoretically could further contract the cloud down into a star. Astronomers have proposed various mechanisms to initiate the process of collapse to the Jeans' length.

One suggested mechanism is the compressive effect of the explosion generated by a nearby supernova. A supernova eruption often expels gas at speeds of more than 600 mi (1000 km) per second. If this wall of moving gas were to collide with a gas cloud, it might compress the cloud sufficiently to trigger star formation. But where did the star that erupted as a supernova come from? Presumably, it formed when a supernova explosion compressed a gas cloud. And where did that star that exploded as a supernova come from? Obviously, this does not explain the ultimate origin of stars, for this theory ends in infinite regress.

Another theory is that dust particles within a gas cloud could act as refrigerators, radiating away heat, thus cooling and contracting the cloud down to its Jeans' length. What often is left unsaid is that formation of dust requires that at least two generations of stars exist before stars can form by this mechanism. All suggested mechanisms to initiate the process of star formation suffer from this chicken and egg problem: there must be stars before stars can form.[13] While these proposed mechanisms might suffice occasionally to form stars today, they do not tell us where stars originally came from. Many astronomers now think that there was a burst of star formation in the very early universe. The mechanism that initiated this is not known, though many astronomers speculate that it may have had something to do with dark matter. Therefore, one may summarize this as "Many stars formed in the early universe by some unknown mechanism." Only to people committed to naturalism would this be considered a scientific statement.

What supposedly happens next as the gas cloud transforms into stars? The *Eddington limit* is the maximum luminosity that a star may have. A star more luminous than the Eddington limit will blow itself apart until its luminosity falls below the limit. For MS stars, more luminous stars are more massive, so the Eddington limit amounts to a maximum mass that a star may have. The exact maximum mass that a star may have is uncertain, but it appears that it is at most a few hundred times the mass of the sun. I must point out that such massive stars are extremely rare, with most stars having a mass less than

---

[13] The old question "Which came first, the chicken or the egg?" is supposed to be an imponderable question. However, the question has a definite answer, depending upon one's worldview. Creationists believe that the chicken came first, while evolutionists believe that the egg came first.

that of the sun. If a cloud of gas has thousands of times the mass of the sun, then it obviously cannot form a single star. Rather, astronomers think that the contracting cloud fragments into many smaller clouds that individually contract to form stars. If this were to happen, the stars that formed would continue to be gravitationally bound. We see many star clusters, collections of stars that are gravitationally bound, so this is taken as evidence that the theory is correct. However, there are many stars, such as the sun, that are not members of any star cluster. Can this be explained? The gravitational interactions between the stars of a cluster can lead to stars escaping. Indeed, some star clusters are so weakly bound gravitationally that they can evaporate. Supposedly, the sun could be an escapee of such a cluster. Or it could be that some stars, including the sun, sometimes form individually.

Supposing that this process of star formation is correct, what happens next? As a cloud, or a portion of a cloud, contracts, it gradually warms and begins to radiate. At first, the temperature is cool by stellar standards, and the radiation mostly is in the infrared part of the spectrum. This protostar is powered by the Kelvin-Helmholtz mechanism, so the lifetime of the protostar is the Kelvin-Helmholtz time. As we saw in the previous chapter, the sun's Kelvin-Helmholtz time is about 30 million years. For more massive stars, the Kelvin-Helmholtz time is shorter, and for stars less massive than the sun, it is longer. Upon the completion of the Kelvin-Helmholtz time, the star reaches the MS on the H-R diagram.[14] At this point, the temperature and pressure in the stellar core is sufficient to initiate and sustain fusion of hydrogen into helium. Nuclear fusion gradually replaces the gravitational contraction as the star settles into the long lifetime possible on the MS.

Is there any evidence that this theory is correct? Astronomers increasingly have offered examples of what they think is evidence, for there often are news stories reporting the birth of a star. Many of these press accounts leave the reader with the impression that astronomers literally saw the birth of a star, as if astronomers beheld a star that was not there a few weeks before. Only by digging deeply into the news story or, even better, consulting the original

---

[14] Caution! Astronomers frequently talk of stars moving on the H-R diagram due to gradual changes in size and surface temperature. Unfortunately, students and other laymen sometimes mistakenly think of this motion as moving through space. The H-R diagram is just that, a diagram illustrating physical properties. Any motion on the diagram is because of changes in physical properties.

source can one come to realize that the situation is far murkier than that. The process of star formation allegedly is very gradual and thus requires much time, so the simple scenario of seeing a star wink on is not correct. What actually has happened in each case is that astronomers, using their theories of how stars form, interpret what amount to snapshots of objects at various stages of stellar formation. For example, stars supposedly form in regions of space in which there are many clouds of gas and associated dust. Stars allegedly form within the interiors of these clouds, but the intervening dust particles block our view. However, infrared radiation can penetrate this dust. Therefore, when infrared telescopes detect point-like sources embedded within these cloud complexes, they are interpreted as forming stars or recently formed stars.

Over the past four decades, the theory of star formation has been honed by comparison with objects that many astronomers think are young. For instance, there are stars in supposed star-forming regions with certain characteristics, such as T Tauri variable stars[15] that have guided this. T Tauri stars exhibit irregular changes in luminosity and have strong chromospheric emission lines in their spectra. Many T Tauri stars have strong stellar winds and probably have many spots on their surfaces. Some of these stars eject gas at a high velocity from their poles. These jets are called bipolar flows, and also are seen in far more energetic situations, such as the cores of active galaxies. The jets probably are caused by strong magnetic fields. The strong magnetic fields can explain other phenomenon of T Tauri stars, such as the spots, winds, and chromospheric emission. Some T Tauri stars have disks of material orbiting in their equatorial planes that are interpreted as planetary systems in formation. Astronomers generally interpret T Tauri stars as proto-stars that are in the last stages of formation just before landing on the MS. This interpretation of T Tauri stars, along with their characteristics, has led the way in refining the theory of star formation.

So have astronomers truly found stars that were just born? If the theory is correct, then perhaps. However, there is much interpretation of the data, interpretation that is heavily laden with assumptions and is very model dependent. A century ago, most astronomers thought that they saw proto-

---

[15] Types of variable stars are named for their prototypes, the first specific star of each type that was studied in-depth. As Cepheid variables are named for the prototype Delta Cephei, T Tauri stars are named for their prototype, T Tauri.

stellar and proto-planetary systems being formed. Photographs of *spiral nebulae* showed round, flattened disks of material with bright, central bulges. These images conformed to ideas at that time of how stars and planetary systems formed from gas clouds, so most astronomers interpreted them as evidence that their theory was correct. In each spiral nebula, the central bulge was viewed as the forming proto-star, and the disk was the material from which the planets would form. While the modern version of this theory is dressed up with far more detail, the basic theory has not changed in more than a century. One astronomer, Adriaan van Maanen, even measured the motion of blobs of material within the disks of some systems, supposedly indicating orbital motion, which is what one would expect if these were planetary systems in formation.

Therefore, it was quite a shock in 1924 when Edwin Hubble showed that these spiral nebulae were not nebulae (clouds of gas and dust) at all, but instead were *entire galaxies* containing billions of stars. Being spiral galaxies, these objects were much farther away than they would be if they were forming stars and planetary systems. The use of the term spiral nebulae to describe spiral galaxies continued for decades afterward, though now it is considered archaic. Being so far away, no orbital motion could be discerned. Therefore, van Maanen's observations were spurious. What went wrong? Van Maanen's expectations were driven by the dominant model of the day, which led him not only to misinterpret data, but to create data that did not even exist. This sorry chapter of astronomical history is almost forgotten today, but it ought not to be, because its lessons are very important. As is the case today, many astronomers a century ago were thoroughly convinced that they had proof of their theory of star formation. And as is also the case today, astronomers then interpreted data and even collected data with the assumption that the theory was correct. Does this parallel prove that astronomers today are completely wrong about star formation as they were a century ago? No, but it ought to teach astronomers to be humbler in their pronouncements on these matters.

Interpretation of the Main Sequence

Stars on the MS generally are stable. As previously mentioned, the MS lifetimes of stars can be very long, at least compared to the Kelvin-Helmholtz time, the theoretical length of time that a star takes to form and reach the MS.

Consequently, astronomers think that stars spend most of their time on the MS. Both the stability and longevity of stars on the MS is due to their deriving their energy from the nuclear fusion of hydrogen into helium in their cores. However, eventually every star will exhaust all the hydrogen fuel in its core. No longer able to sustain itself by this nuclear source, a star cannot remain on the MS, so what happens next? To answer this question, astronomers have developed an elaborate theory of stellar evolution, which is the subject of the next chapter. In the remainder of this chapter, I will discuss the MS.

In the previous chapter, I discussed the faint young sun paradox. Here I briefly will describe the physics behind the faint young sun paradox. As MS stars consume the hydrogen fuel in their cores, they replace it with helium, the byproduct of the nuclear reactions. This reduces the number of particles in stellar cores, or, alternately, the mean molecular weight of the matter in the cores increases. The ideal gas law[16] adequately describes the behavior of the gas in stellar cores. As the number of particles decreases, the ideal gas law requires that the other variables change as well. More specifically, the temperature and pressure in the core increase while the volume decreases. The nuclear reactions that power stars are very temperature sensitive (and are somewhat less sensitive to pressure). As the temperature increases, the rate of nuclear fusion increases. Therefore, as a star ages on the MS, its energy production slowly increases. The increased luminosity is driven outward through the *envelope*, the region between a star's core and its photosphere. The increased energy passing through the envelope generally expands the envelope, enlarging the star. This expansion can lead to a very slight cooling of the star's photosphere. In most cases, the cooling is modest, so the larger size increases the star's luminosity.[17] Therefore, astronomers expect stars on the MS to move slightly upward and a little to the right on the H-R diagram as they age. This is consistent with the MS being a band of stars with some vertical thickness rather than a thin line. In addition to age, some of the thickness of the MS is explainable in terms of different initial compositions of stars.

---

[16] The ideal gas law has the form $PV = nRT$, where P is the pressure, V is the volume, n is the number of particles, usually expressed in moles, R is the ideal gas constant, and T is the temperature.

[17] There is a balance between the energy a star produces and its luminosity. This principle is called *thermal equilibrium*.

The previous chapter mentioned that the sun's MS lifetime is approximately 10 billion years. This is based upon the assumption that about 10% of the sun's mass is in its core, where the thermonuclear reactions that power the sun occur. From spectroscopic analysis, we know that the sun's photosphere is 75% hydrogen. If the sun formed from a thoroughly mixed gas cloud, then the rest of the sun, including its core, originally was 75% hydrogen as well. The hydrogen fusion reactions that power the sun convert 0.7% of the mass into energy via Einstein's mass-energy equivalence equation, $E = mc^2$. Multiplying these percentages with the sun's mass and dividing by the sun's luminosity yields the solar lifetime of 10 billion years. Presumably, similar estimates apply to other stars as well. Rather than compute this for each star individually by plugging in its mass and luminosity, we could scale this using solar units. Expressing stellar mass and luminosity in terms of solar mass and solar luminosity (where the sun's mass is one, and the sun's luminosity is one), the lifetime of a star, expressed in solar lifetimes (one solar lifetime is 10 billion years), we get:

$$T = M/L.$$

Inserting the mass-luminosity relation, this becomes,

$$T = M^{-2.5},$$

from which we can see that high-mass stars have the shortest MS lifetimes, while low-mass stars have the longest MS lifetimes.

The lowest mass for a star on the MS is about 7% the mass of the sun. From this equation, the MS lifetime of such a star would exceed a trillion years. However, if one uses a more appropriate exponent in the mass-luminosity relation for the lower MS, the lifetime turns out to be hundreds of billions of years. At any rate, the MS lifetime of the lowest mass stars greatly exceeds the current 13.8-billion-year age estimate of the big bang universe, rendering the fate of low-mass stars a moot point. However, the situation is very different for high-mass stars. A star that is ten times more massive than the sun has an MS lifetime of about 30 million years, while a star with 50 solar masses has an MS lifetime of less than a million years. These are only estimates, but they clearly show that high-mass stars have extremely short lifetimes compared to the supposed billions-of-years age of the universe. How do evolutionary astronomers explain this? They argue that stars continually form, replacing older stars as they age and cease to exist. Some biblical creationists have used the fast burn rate of massive stars as evidence that the universe is young, but this is not a good argument, if stars do indeed form. Therefore, the fast burn rate of massive stars amounts to an argument against star formation.

Since energy generation by fusion of hydrogen into helium in their cores characterizes MS stars, once a star exhausts its hydrogen in its core, it no longer can remain on the MS. Therefore, astronomers expect stars to move off the MS on the H-R diagram after its MS lifetime. This leads to current theories of stellar evolution, the topic of the next chapter.

CHAPTER 8

Differing From One Another in Glory:
Explaining Stellar Diversity

In the previous chapter, I briefly surveyed what we know about stars. I also discussed the structure of stars, the source of stellar energy, and our understanding of what the main sequence (MS) is. I ventured into a more speculative realm in the treatment of how many astronomers think stars naturally form. In this chapter, I will continue with more uncertain ideas of how stars of advanced age evolve. From time to time, I will make comments about these theories in the context of biblical creation. Much work on a biblical theory of stars remains to be done.

Post Main Sequence Evolution

Astronomers think that MS stars derive their energy from the fusion of hydrogen into helium in their cores. What do astronomers think happens to an MS star once it exhausts all the hydrogen fuel in its core? With no more nuclear fuel, the core must shrink, tapping gravitational potential energy as an energy source. This is the Kelvin-Helmholtz mechanism once again, but it kicks in only briefly. As the core shrinks, its temperature increases. The temperature in a thin shell in the envelope surrounding the core also increases. The temperature increase in the thin shell initiates fusion of hydrogen into helium in the thin shell, and this becomes the star's primary energy source. Since the temperature now in the shell is higher than the core temperature previously had been, the nuclear reaction rate increases over what it had been in the core. Therefore, more energy must pass through the rest of the envelope to the star's photosphere. This increase in energy transport heats the envelope, which expands the gas in the envelope, enlarging the star. As the outer layers of the star expands, it cools. It is a bit paradoxical, but as a star's core temperature increases, its photospheric temperature decreases. On the Hertzsprung-Russell (H-R) diagram, the star moves off the MS to the upper right, becoming a red giant. As the star ascends higher in the red giant region, its core continues gradually to shrink and heat.

The theoretical understanding of what happens next depends upon a star's mass. Consider the sun and other stars with similar mass. If the sun is 4.6 billion years old, then it has exhausted about half of its 10 billion-year MS lifetime. Therefore, in about another 5 billion years, most astronomers expect the sun to depart the MS. As a star like the sun ascends the red giant branch, its core's mass increases as more helium from fusion in the shell around the core is added to it. Meanwhile, the core slowly contracts and heats. Eventually, the core temperature gets hot enough to initiate fusion of helium into carbon. The energy released by this process abruptly heats the core, resulting in a brief runaway, an event that astronomers call the *helium flash*. Most of the energy generated is absorbed by the core, expanding it. This tends to regulate the helium fusion, which leads to stable energy generation in the core. With an expanded and reinvigorated energy source in the core, the envelope of the star shrinks. On the H-R diagram, the star moves downward toward the MS, but it does not quite reach the MS. Instead the star settles onto the *horizontal branch*, so called, because on the H-R diagram, these stars lie along a horizontal line above the MS. The horizontal branch is a relatively stable time in a star's development, though not nearly as long-lived as the MS.

Eventually, all the helium in the core is fused into carbon, so helium fusion ceases in the core. Once again, the star hypothetically will ascend in the H-R diagram to become a red giant. To distinguish it from the first time it was a red giant, astronomers call this the *asymptotic giant branch*. A giant branch star's source of energy is fusion of hydrogen into helium in a thin shell around a helium core. However, an asymptotic giant branch star is a double shell source star. That is, two different nuclear reactions occur in thin shells around a carbon core. The outer shell fuses hydrogen into helium. Around the star's core, this helium is further fused into carbon. Astronomers think that an asymptotic giant star eventually will develop powerful winds that strip off its outer layers. Once all the envelope is removed, all that remains will be the core, which we would now recognize as a white dwarf. Of course, if this is the future of the sun, long before this happens, the intense light of the expanding red giant sun would fry the earth or possibly even engulf it.

The gas that this process expels piles up to form a *planetary nebula* around the now naked core. The term planetary nebula is a misnomer, because planetary nebulae have nothing to do with planets. They were called this a couple of centuries ago because the first ones discovered appeared round through the largest telescopes then available, and thus superficially resembled planets. For some time, astronomers thought that planetary nebulae where spherical shells of gas, but now they think that most planetary nebulae are hourglass-shaped. Depending upon their orientation, planetary nebula can appear very differently, such as round or a cat's eye (Images G, H, I, J, and K). Since they are comprised of gas expelled from stars, many people mistakenly think that planetary nebulae were blown off by supernovae explosions, but those events produce something quite different. Radiation from the central star causes the gas to glow so that we can see a planetary nebula. As the star cools and the gas expands and dissipates, the planetary nebula disappears, leaving behind just the white dwarf. Thus, this theory explains how both planetary nebulae and white dwarfs form. The fact that every planetary nebula has at its center a very hot star resembling a white dwarf is taken as evidence that this theory is correct. While every planetary nebula has such a star at its center, most white dwarfs do not have a surrounding planetary nebula. The explanation for this is that planetary nebulae are very short-lived, existing perhaps for tens of thousands of years, while white dwarfs can exist for many billions of years. White dwarfs slowly cool as their immense stored thermal energy is tapped to power them.

This is the general understanding of the sun and stars with similar mass, but what about other stars? Lower mass stars probably will ascend the red giant branch the first time, but they likely do not possess enough mass to ignite the nuclear reactions that produce the asymptotic giant branch, and perhaps not even the horizontal branch. Instead, they would develop strong winds that would strip away the envelope to produce a white dwarf from their cores. The question of the fate of the lowest mass stars is a moot point, because most astronomers think that the universe has been around for less than 14 billion years, far less time than the MS lifetime of the lowest mass stars.

The fate of high-mass stars gets more interesting. More massive stars appear to be capable of core temperatures and pressures that can support nuclear reactions beyond the fusion of helium into carbon. Therefore, more massive stars probably can move back and forth across the upper regions of the H-R diagram several times as new nuclear fuels are exploited, only to be exhausted eventually. All the while, the stellar cores become progressively hotter and denser. However, there is a limit to how much matter can accumulate in the cores of these stars. When the cores of these stars exceed this limit, the cores catastrophically collapse into what astronomers call *compact objects*. Compact objects can be of two types: *neutron stars* and *black holes*. I will discuss these shortly, but suffice it to say for now that they are about a thousand times smaller than white dwarf stars, yet contain more mass. The collapse of stellar cores into compact objects releases a tremendous amount of energy. This energy can exceed the entire amount of energy that the sun would emit over its MS lifetime of 10 billion years. The difference is that this energy is released in an instant. This powerful event rips through the envelope of the star. This process takes about a day, and upon reaching the photosphere, the star is disrupted in a titanic explosion called a *supernova*.[1]

How much of the process of post-MS stellar evolution has been observed? Except for supernovae, none of it. Instead, as with star formation, astronomers have what amount to snapshots of various objects, such as white dwarfs and red giant stars, that they have assembled into the framework of how astronomers think these objects came to be. That is, the mere existence of these various types of stars does not prove the evolutionary theory, but rather the evolutionary theory is used to interpret how these various stars interrelate. Even supernovae, which we do see abruptly happen, cannot be taken as proof of this theory either, for they, too, have been interpreted in terms of the evolutionary paradigm.

When asked for evidence that their theories of stellar evolution are true, astronomers usually turn to the H-R diagrams of star clusters. A star cluster is a large group of stars that is gravitationally bound. One of the best examples of

---

[1] Besides the large difference in energy involved, supernovae and novae are very different events. Novae are recurring temporary increases in brightness in close binary stars fueled by mass transferred from one star to the other. Supernovae are powerful events that occur once, transforming, or even destroying, the stars involved.

a star cluster is the Pleiades.[2] There are two types of star clusters, open clusters and globular clusters. As the name suggests, globular star clusters are spherical, but open star clusters have irregular shapes. Globular clusters are much more rich than open clusters, typically containing between 50,000 and a half-million stars, while open clusters have hundreds or at most a few thousand stars. It is extremely unlikely that random stars could be assembled into clusters, so it seems obvious that stars in a cluster must have formed at the same time, probably from a single gas cloud. Assuming that the matter in the gas cloud was mixed so that the cloud had about the same composition throughout, stars within a cluster ought to have the same initial composition; and since the stars formed at the same time, they ought to have the same age as well. Therefore, any differences between stars within a cluster ought to be the result of different masses.

As stars use up their stores of energy, on the H-R diagram they ought to move slightly upward and a little to the right across the MS band before departing the MS toward the red giant region. Since more massive stars have shorter MS lifetimes, we would expect them to evolve first. Because the more massive stars are at the top of the MS, this evolution ought to show up as a truncation of the MS on the upper end with a little turning upward to the right at that truncation (Figure 8.1). Of course, this process would take so long that we could never watch it in real time. However, all the stars within a particular star cluster ought to have approximately the same age, but different clusters probably would have different ages. Therefore, comparison of H-R diagrams of different star clusters[3] ought to reveal differences due to age (Figure 8.2). When we examine the H-R diagrams of star clusters, we normally can pick out the MS, and, as expected, MS terminates on the upper end at some point. If the cluster has enough stars, the stars on the MS appear to turn away and head up toward the red giants. Astronomers call this the *turn-off point*. Each star cluster has a different turn-off point. Clusters with higher turn-off points are interpreted as being younger, while clusters with lower turn-off points are older.

---

[2] A photograph of the Pleiades appears on the front cover of the companion book, *The Created Cosmos: What the Bible Reveals about Astronomy* (Danny R. Faulkner, Green Forest, Arkansas: Master Books, 2016).

[3] The *observed* H-R diagrams of star clusters usually are color-magnitude diagrams, so called, because the observations consist of magnitudes versus color differences of magnitudes, such as B-V.

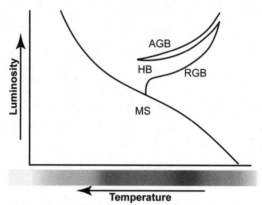

Figure. 8.1. H-R diagram highlighting massive MS stars.

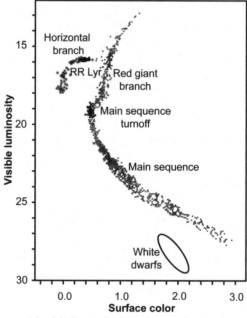

Figure 8.2. H-R diagram highlighting different turnoff points.

The following figure shows a comparison of schematic observed H-R diagrams of the two types of star clusters (Figure 8.3). Notice that the turn-off point of the open cluster is much higher, indicating younger age. There is quite a range in the heights of the turn-off points of open clusters, which is interpreted as a range in ages. The Pleiades star cluster has a relatively high

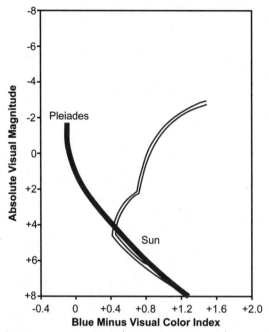

Figure 8.3. Composite H-R diagram (Pleiades and M3).

turn-off point; its age is fixed at less than 100 million years. Even younger still are the star clusters h Persei and χ Persei, often referred to collectively as the *Double Cluster*. The two clusters have the same inferred age, little more than 10 million years. On the other extreme, the open cluster NGC 188 has an inferred age of 6.8 billion years. There is far less range in the turn-off points of globular clusters, indicating less diversity in age. Globular cluster ages typically are thought to be 12 billion years. Notice that on the H-R diagram of the globular cluster there is a horizontal region above the MS. Astronomers call this the *horizontal branch*. Theory suggests that stars on the horizontal branch derive their energy from fusion from helium into carbon in their cores. Their location on the H-R diagram gives the name to these stars, horizontal branch stars.

While this concordance between the H-R diagrams of star clusters and stellar evolution theory often is put forth as evidence for the theory of stellar evolution, it is not actually proof. Rather, it is the way observations are interpreted in terms of the current models of stellar evolution. One cannot interpret data in terms of a model and then use that interpretation to prove that the model is true. Furthermore, the detailed observed H-R diagrams of many star clusters became available in the early 1950s, just before the modern theory of stellar structure and evolution was developed in the mid and late 1950s. Already, astronomers had decided that globular clusters were older and open clusters were younger. Therefore, the very observations that supposedly are predictions of the theory were in hand and used as a guide in developing the theory. To claim that this interpretation of the data proves the theory is circular reasoning.

Sometimes biblical creationists point out a few anomalies which suggest that the current theories of stellar evolution are wrong. One anomaly is the existence of *blue stragglers*. A blue straggler is a member of a star cluster lying on the MS, but beyond (toward the left, or blue, end of the horizontal axis) the cluster's turn-off point. Blue stragglers exist in some, but not all, globular clusters, as well as some of the older open clusters. The MS lifetime of a blue straggler is much younger than the inferred age of the star cluster, so why has this star not evolved off the MS already? This has been a perplexing problem for stellar evolution for decades, and there is no clearly agreed upon answer to it.

Some biblical creationists have brought up the break up time of star clusters as evidence of recent origin. Open star clusters are loosely bound. The gravitational attraction of other objects in the galaxy pulls on individual members of star clusters, gradually tearing the clusters apart. For instance, dynamical studies of the Pleiades indicate that it will break up within 250 million years, so it must not be very old. Indeed, as discussed above, astronomers think that the Pleiades is less than 100 million years old, so this is not a problem. Many astronomers think that the sun formed in a star cluster, but that star cluster long ago evaporated, leaving the sun as a single star.

Biblical creationists also cite examples of individual stars that supposedly defy theories of stellar evolution. For instance, in the *Almagest*, the second century AD astronomer Claudius Ptolemy recorded that Sirius, the brightest appearing star in the sky, was red. Today it is blue-white. Stellar evolution would not allow for Sirius to change color so dramatically and so quickly but, more particularly, it would not change from red to blue-white. To some creationists, this suggests that stellar evolution does not proceed in a manner anything like the current theory. Other discussion has centered on Sirius B, a white dwarf that orbits Sirius A.[4] Being an end-state of stellar evolution, presumably Sirius B was a red giant star sometime in the past. Some biblical creationists think that this is evidence that stellar evolution happens very rapidly. However, it would have been difficult for Sirius B to have blown off its outer layers and become a white dwarf so quickly and not leave any evidence, such as a planetary nebula, or at least some noticeable circumstellar gas. How do secular astronomers explain this odd color anomaly with Sirius? They note that other ancient sources record the same color that we see today. This raises the possibility that Ptolemy may have been referring to something else other than color when he wrote that Sirius was red. In ancient cultures, a few stars and constellations were associated with certain things, such as fire or water. It could be that his description of Sirius as red referred to Sirius' association with fire rather than its true color.

FG Sagittae (FG Sge) is another star that biblical creationists occasionally mention in the context of stellar evolution. In 1955, astronomers determined that FG Sge's spectral type was B. In 1991, only 36 years later, its spectral type was G, but eventually it became a K-type star. This is a drastic change in color from blue, to yellow, to orange. Within the theory of stellar evolution, stars generally do not change this quickly; hence, some biblical creationists use the case of FG Sge to argue that there is something wrong with that theory. However, this puts creationists in the peculiar situation of saying that stellar evolution does not happen, but when it does, it happens very quickly. Over the years, astronomers have learned much more about FG Sge, and they have found other stars that resemble it. Most notable are the stars V4334 Sagittarii (V 4334 Sgr, also known as Sakurai's object) and V605 Aquilae (V605 Aql). All three of these stars are central stars of planetary nebulae. Astronomers believe that these stars are on their way to becoming white dwarfs, but have not reached that status quite yet. The immediate precursors of white dwarf stars probably are asymptotic giant branch stars. Astronomers expect that these stars undergo *thermal pulses*, episodic occurrences of thermonuclear reactions in the layers surrounding their cores. Astronomers think that very late thermal pulses briefly can rejuvenate a dying star before it finally becomes a white dwarf. These three stars are thought to be undergoing late thermal pulses.

---

[4] The capital letter designations refer to the members of a binary system. The brighter star is the A component, and B component is the fainter component. Being more than 10,000 times brighter, the light of Sirius A dominates the light that we see with the naked eye.

Thermal pulses in asymptotic giant branch stars have been invoked to explain unusual characteristics of some red giants. Carbon stars are red giants that have an overabundance of carbon. Normally, oxygen is more abundant than carbon in stars, but the situation is reversed in carbon stars. Astronomers think that carbon was produced in thermal pulses in these stars, followed by convection that dredged the carbon up to the stars' photospheres. Metal stars, stars with unusually high abundances of metals with atomic numbers greater than iron, such as zirconium, yttrium, strontium, barium, and lanthanum, are similarly explained. The theory is that these metals recently were synthesized in thermal pulses and dredged up to the photospheres of these stars. Some metal stars have absorption lines of technetium in their spectra. There are no stable isotopes of technetium, and none of its isotopes have half-lives more than a few million years. Some biblical creationists have pointed out that the presence of technetium in these stars shows that the stars are not billions of years old. However, no one suggests that the technetium has been present in these stars for nearly that long. Instead, the technetium was recently introduced and eventually will decay away.

The End States of Stellar Evolution:
White Dwarfs, Neutron Stars, and Black Holes

Earlier in this chapter I briefly introduced white dwarfs. They are approximately the size of the earth, yet have mass comparable to the sun. Combining the mass and size, we can calculate the densities of white dwarfs. White dwarf density is very high; in some cases the density is about a million times that of water. On earth, the densest substances are a little more than 20 times the density of water. When astronomers discovered white dwarfs and soon learned their density in the late nineteenth century, the structure of white dwarfs was a profound mystery. Stars are subject to hydrostatic equilibrium, a well-understood principle with myriad applications. For stars, hydrostatic equilibrium is the balance between the inward force of gravity and the outward force of gas pressure. Knowing the gross properties of white dwarfs, astronomers could compute the surface gravity and gas pressure within white dwarfs, and they found that gas pressure was woefully inadequate to overcome the force of gravity. Therefore, white dwarfs ought to rapidly contract, yet they obviously were very stable stars.

The 1920s saw the development of quantum mechanics, and soon astrophysicists[5] began to apply quantum mechanics to the structure of white dwarf stars. Particularly noteworthy was the work of Subrahmanyan Chandrasekhar. In the interior of most stars, the gas is totally ionized. According to quantum mechanics, electrons cannot be degenerate, which means that two or more electrons cannot occupy the same energy state. Within the interiors of most stars, there are plenty of energy states available, so this is not a problem. However, in the extremely high density of a white dwarf, all the available energy states are occupied, so if the white dwarf were compressed any more, electrons would become degenerate. Since electrons are prohibited from being degenerate, they strongly resist any further compression. This effectively amounts to a pressure, so astrophysicists call this *electron degeneracy pressure*. It is electron degeneracy pressure which provides the bulk of the pressure that opposes the contraction of gravity in white dwarf stars.

While the physics upon which our understanding of white dwarfs is based may be high-powered, the structure of white dwarfs is relatively simple, so our understanding of white dwarf structure is very good. We may understand white dwarfs better than we understand the sun. The theory of white dwarf structure makes a remarkable prediction: there is a maximum mass that a white dwarf may have and be supported by electron degeneracy pressure. The exact value of this *Chandrasekhar limit* depends upon composition, but the approximate value is about 1.4 times the mass of the sun. White dwarfs are commonly found in binary stars, so we have a large sample of masses of white dwarf stars. White dwarf masses range from less than half a solar mass to 1.4 solar masses, with a large clump of stars near the 1.4 solar mass limit, but none beyond. This is very good evidence that our theory of white dwarf stars is correct.

What if the mass of the core of a star exceeds the Chandrasekhar limit? Electron degeneracy pressure would not be sufficient to balance gravity, so the

---

[5] The term *astrophysics* was coined in the late nineteenth century to describe a new emphasis in astronomy on chemistry and physics, primarily as it related to the study of stellar spectra. In the early twentieth century, modern physics, with its twin pillars of quantum mechanics and general relativity, became part of the mix. *Astrophysicist* is a term that some astronomers or some physicists who specialize in astronomy prefer, but there is no clear distinction between astronomers and astrophysicists. In some cases, as is the use here, it is to emphasize the use of modern physics in a theoretical sense rather than emphasizing the presentation of astronomical observations.

core would begin to contract. To prevent degeneracy, electrons quickly invade the nuclei of atoms, where they combine with the protons to form neutrons. With electrons no longer present, electron degeneracy pressure abruptly is removed, and the core almost instantly collapses into a compact object. The gravitational potential energy released is the source of energy that powers the subsequent supernova explosion. The compact object left behind could be a neutron star, a sphere of neutrons only a few miles across but containing tremendous mass—up to three times that of the sun. The collapse stops at this point because, like electrons, neutrons cannot be degenerate either, so a sort of neutron degeneracy pressure prevents further contraction of the neutron star. The density of a neutron star is millions of times greater than the density of a white dwarf. How can any object be this dense? The neutrons of a neutron star are all packed together as tightly as protons and neutrons are in the nucleus of an atom. Normally matter mostly is empty space, with great distances between the nuclei of atoms and their orbiting electrons, and even greater distances between adjacent atoms. But the density of nuclei and neutron stars is extremely high and approximately the same.

As you may have anticipated, the masses of neutron stars are subject to an upper limit just as white dwarfs are. The exact value of this limit is not known, but it appears to be close to three solar masses. What if the mass of a compact object exceeds this mass limit? There is no known remaining force that can balance the pull of gravity, so the compact object presumably would collapse into a black hole. To understand what a black hole is, we must consider a few other things.

Given the mass and radius of an object held together by gravity, we can calculate the *escape velocity*, the vertical velocity required to launch from the object's surface and escape its gravity. For the earth, the escape speed is 7 mi per second (11 km per second), or 25,000 mi per hour (40,000 km per hour). For a black hole, the escape velocity is the speed of light, so not even light can escape from a black hole. The radius from the center where the escape velocity is equal to the speed of light is the *Schwarzschild radius*, which defines the size of a black hole. The equation for the Schwarzschild radius is

$$R_S = 2GM/c^2,$$

where R_S is the Schwarzschild radius, G is the gravitational constant, M is the mass of the body, and c is the speed of light. The Schwarzschild radius for an object with a mass equal to that of the earth is 9 mm (about a third of an inch), while the Schwarzschild radius of the sun is 3 km (less than 2 mi).

The physics that we use to describe black holes completely breaks down within the Schwarzschild radius, so we have no idea what is going on inside them. Perhaps one day we will develop the mathematics and the physical models that adequately describe the interiors of black holes, but for now it remains a mystery. However, our science does work outside of the Schwarzschild radius, so we know that the influence of the gravity of black holes operates normally there. Since we cannot see black holes, you may wonder how we can detect them. We know that black holes exist because of their gravitational influence.

Consider a close binary star, where one of the orbiting objects, or components, is a black hole, but the other component is a normal star. If the two orbit one another closely enough, the tidal force of the black hole will lift matter from the other star, causing it to fall toward the black hole. This phenomenon of mass transfer is well documented in close binary stars where the two stars involved are normal stars. Due to constraints of angular momentum, the transferring matter cannot fall directly onto the black hole. Instead, the transferring matter accumulates in a disk of material orbiting above the Schwarzschild radius of the black hole. Astronomers call this disk of material an accretion disk; and again, accretion disks often are observed in close binaries where both stars are normal stars. Due to frictional losses of energy within the accretion disk, disk material slowly spirals downward toward the Schwarzschild radius, so material gradually feeds into the black hole, increasing its mass. As matter falls downward, it converts a tremendous amount of gravitational potential energy into kinetic energy. Friction in turn converts that kinetic energy into heat, causing the accretion disk to reach incredibly high temperatures, more than 1,000,000 °K. Given their high temperatures, accretion disks around black holes emit copious amounts of X-rays. Since the 1960s, NASA has launched X-ray telescopes into space, and they have detected many X-ray sources, including binary stars. We call such binary stars *X-ray binaries.*

Because only compact objects have the tremendous gravitational potential energy necessary to produce X-rays in such large amounts, one of the components of an X-ray binary must be a compact object. However, how do we know if the compact object is a neutron star or a black hole? Being a binary star, we can measure the mass of the system. If we can infer the mass of the normal star from such things as its spectral type, then we can subtract its mass from the total mass to determine the mass of the compact object. If the mass of the compact object is less than the maximum mass for a neutron star, then the compact object probably is a neutron star. However, if the mass of the compact object exceeds the greatest possible mass of a neutron star, then it must be a black hole. There are hundreds of X-ray binaries known, many with sufficient information so that we can conclude what sort of compact objects they contain. An example of an X-ray binary with a neutron is Scorpius X-1, the first extrasolar X-ray source discovered. It contains a minimal mass neutron star, about 1.4 solar masses. On the other hand, the compact object in Cygnus X-1, another X-ray binary discovered early, appears to have a mass about 15 times that of the sun, making it a black hole.

There is no reason to think that all compact objects exist in close binary systems. There is a way to discover isolated neutron stars. All stars rotate, and all stars have magnetic fields. Being the collapsed cores of massive stars, neutron stars must spin and have magnetic fields too. Since they are much smaller than their progenitor stars, neutron star spin rates are very fast, and their magnetic fields are very strong. A neutron star's intense magnetic field rapidly spins as the star rotates. Thus, charged particles near the neutron star experience a magnetic field that is moving at a very high speed, close to the speed of light. Relative motion between charged particles and a magnetic field accelerates the charges (this is how an electrical generator works). Accelerated charges radiate energy (this is how an antenna on a broadcast radio works). Because of the extremely high velocities and magnetic fields involved, the amount of energy that these accelerated charges produce is immense. Nearly all this radiation is beamed along the axis of the neutron star's magnetic field. Magnetic fields and rotation axes rarely coincide, so the magnetic field, and hence the beam of radiation, sweeps out a cone-like a searchlight. If we happen to lie near the swept out cone, we observe periodic flashes of radiation. Most of the radiation is in the radio part of the spectrum.

The first observations of this phenomenon were by Jocelyn Bell (Burnell) in 1967, consisting of periodic pulses of radiation every 1.33 secs coming from a point source in the sky. At first, astronomers did not know what to make of this. Their first thoughts were that it was some sort of interference or that it was from a terrestrial source; but both of those possibilities were eliminated quickly. The precise regularity suggested a beacon, so they half-jokingly considered the possibility that it might be an alien radio transmission (they used the acronym LGM, for *little green men*, to describe this possibility). They soon concluded that it was a natural source, and they called it a *pulsar*, a portmanteau of *pulsating star*.[6] Several astronomers quickly suggested the explanation of a rapidly rotating neutron star, as described above (neutron stars had been theorized much earlier, in 1934). Soon other pulsars were discovered. Their periods range from a few seconds down to milliseconds. Today, there are more than 2000 known pulsars. In the case of most pulsars, we would not expect that we lie close enough to the swept-out cone so that we detect the radio pulses. Thus, the number of pulsars that we will never detect must dwarf the number that we can detect. Astronomers estimate that there must be at least a few hundred thousand pulsars in our galaxy.

Can we be sure that our interpretation of pulsars as rapidly rotating neutron stars is correct? There are reasons for confidence. First, the theory makes specific predictions about the radiation that we detect. If it is generated in the manner described here, then the radiation ought to have a very characteristic spectrum called a *synchrotron spectrum*. Furthermore, the electromagnetic waves making up the radiation ought to vibrate a certain way. This phenomenon is called *polarization*. The radiation coming from pulsars have both a synchrotron spectrum and the proper polarization. If that were not enough, some compact objects in X-ray binaries happen to be pulsars as well, confirming the identity of the compact objects involved as neutron stars. Mass estimates of the compact objects in these systems confirm that these are neutron stars and not black holes. Finally, several pulsars are associated with supernova remnants, the detectable remains of a supernova thousands of years after the supernovae occurred. In most cases, the supernova remnant only is inferred to be from a supernova explosion, because the explosion happened so long ago that there are no historical records of the event. A notable exception, however, is the supernova that happened on July 4, 1054. Its location coincides

---

[6] This name is a bit of a misnomer, because it might suggest to some people that pulsars pulsate in size. Pulsars do not change size, but rather their spinning motion causes their output to pulsate from our perspective.

with that of the Crab Nebula. The expansion of the Crab Nebula indicates an age of nearly 1000 years, consistent with the known age of the supernova. Near the center of the Crab Nebula is the Crab Pulsar (PSR B0531+21), one of the first pulsars discovered. While pulsars are amazingly regular in their pulses, they do slow down very gradually. This is expected, as the radiation taps the kinetic energy of the rotating neutron star, so its rotation must slow as the radiation is emitted. Astronomers use the spin-down rates of pulsars to estimate their ages. The Crab Pulsar's age is consistent with the known age of the supernova. Not only does all this interrelated evidence give us confidence of the existence of neutron stars, but it also indicates that astronomers likely are on the right track in interpreting the end of at least the more massive stars.

The eruption of a supernova produces an expanding cloud of debris that can emit radiation for many years after the eruption. The Crab Nebula is an example of one of these *supernova remnants*. We have a good understanding of the physics of supernova remnants, so we can expect supernova remnants to exist for some considerable time. The creationist Keith Davies noticed a profound lack of old supernova remnants,[7] suggesting that this is a good indication of recent origin. While this looks promising, it relies upon a correct understanding of how supernova remnants evolve. If the later stages of supernova remnants are not properly understood, then this indicator of recent creation may be suspect.

Many biblical creationists express doubts about the reality of black holes. We shall encounter far more massive black holes lurking in the cores of galaxies in the next chapter. There is much evidence for those supermassive black holes, as well as for stellar black holes, discussed in this chapter. So why do some creationists doubt their existence? The answer probably relates to a confusion between operational science and historical science, as discussed in the Introduction. While there is a good case for the existence of black holes from operational science, much of the discussion of black holes unfortunately is in the context of proposed evolutionary history. From reading these accounts, it is easy to get the wrong impression that black holes were made up by astronomers somehow to support their evolutionary ideas. The observations that support

---

[7] Keith Davies, "Distribution of Supernova Remnants in the Galaxy," *Proceedings of the Third International Conference on Creationism*, edited by R. E. Walsh (Pittsburgh, Pennsylvania: Creation Science Fellowship, 1994), 175–184.

the existence of black holes came first, but once astronomers came to realize that black holes exist, they felt compelled to tell the story of black holes within the paradigm of evolution. There is no good reason for creationists to doubt the existence of black holes.

Extrasolar Planets

Extrasolar planets are planets orbiting other stars. Four centuries ago, people began to abandon the geocentric theory in favor of the heliocentric theory, the idea that the earth is one of several planets orbiting the sun. It soon followed that the stars are very far away and in many respects like the sun. This suggested the possibility that other stars have orbiting planets, just as the sun does. Intimately related to this is the belief in *plurality of worlds*—that other planets support life, including beings like humans. Belief in the plurality of worlds had become very popular by the early nineteenth century, with many people thinking that life existed on all the planets, on the satellites of planets (such as the moon), and even on the sun and other stars. This exuberance has waned considerably since then, but belief in life on other planets persists today. Life is not likely on any of the other planets in the solar system, so life elsewhere, if it exists at all, must be on planets orbiting other stars, hence the current interest in extrasolar planets. Much of this expectation is based upon the assumption of evolution. If life developed naturally on earth, then one would expect that life developed on other planets as well. Otherwise, the earth would be unique, and that would imply the earth holds a special status. Evolutionists avoid admitting any special status for the earth, because that would suggest creation. Biblical creationists generally see this connection as well, and so many creationists mistakenly think that the existence of extrasolar planets somehow undermines the creationary worldview. As we shall see, the exact opposite is the case.

For a long time, astronomers speculated on the existence of extrasolar planets, and a few even thought that they had evidence of planets orbiting other stars.[8] There are several reasons why detection of extrasolar planets is so difficult. First of all, direct detection of extrasolar planets is very difficult,

---

[8] The best example is Peter van de Kamp, who, starting in 1960, published papers claiming a wobble in the movement of Barnard's Star, the fourth closest star to the sun. He thought that this wobble was due to the influence of a massive orbiting planet. Most astronomers consider his results to be spurious.

because planets are much fainter than the stars that they orbit. Even for a nearby star, any orbiting planets would appear close to the star, and thus are likely to be lost in the glare of the star. More promising is indirect detection of extrasolar planets. One method of indirect detection is to measure the motion of the star as the result of the gravity of an orbiting planet. Because of Newton's third law of motion, as a planet orbits a star, the star also orbits the planet. Of course, since the star has far more mass, it does not move nearly as much as the planet does. Still, there ought to be a slight periodic Doppler effect in the spectrum of a star with an orbiting planet or planets. However, this Doppler signature is very feeble and easily overwhelmed by other effects on the spectrum. Another indirect method of detecting extrasolar planets is to observe the transits[9] of planets as they pass in front of their stars once each orbit. For instance, for a suitably located observer looking toward the sun, the earth would transit the sun for about 13 hours once per year, dimming the sun by nearly $1/10{,}000$. Jupiter's transits would last longer and would dim the sun by $1/100$, but those eclipses would occur nearly a dozen years apart. Obviously, such precise measurements of such infrequent and generally unpredictable times[10] would be very difficult to make. Furthermore, transits of extrasolar planets can be observed only if we lie close to the orbital plane of those planets. This is the case only for a small minority of extrasolar planets, so we can discover at most only a small fraction of all extrasolar planets this way.

   With these difficulties, detection of extrasolar planets was not possible until the 1990s, when advances in technology made it much easier. For instance, improvements in computer software made it possible to ferret out the weak periodic Doppler motion induced by an orbiting planet among all the other effects in the spectra of stars. Improvements in the sensitivity of observations permitted direct imaging of a few planets, and further advancements soon will aid direct detection of many more extrasolar planets. One of the greatest advancements in extrasolar planet discovery has been the automation of telescopes and improved precision of measurements that allowed the detection

---

[9] A transit is a sort of eclipse, where the eclipsing body appears smaller than the eclipsed body. An occultation is the reverse—when an apparently larger body eclipses a smaller body.

[10] It is not that the transits could not be predicted. They could be predicted, once we knew that the planet existed and understood its orbital parameters. But prior to discovery, we cannot anticipate when any particular extrasolar planet would transit its star.

of many extrasolar planets via transits. However, the greatest single boon to the detection of extrasolar planets was the launch of the Kepler spacecraft in 2009. The main part of the mission lasted four years, but a limited program continues. Kepler continually monitored the brightness of nearly 150,000 stars with a precision of one part in a million. As of 2016, Kepler had discovered nearly 1300 extrasolar planets. At the time of the writing of this book, the total number of known extrasolar planets from all programs is nearly 4000.

The search for extrasolar planets has three primary goals:

1. To show that planets are common.
2. To show that planetary systems are common.
3. To show that earth-like planets are common.

Given that we have found nearly 4000 extrasolar planets thus far, and that the discovery of new ones continues, it seems that planets are abundant. We have found examples of many stars with multiple planets, so planetary systems probably are common. But what about earth-like planets? To answer that, we must ask what an earth-like planet is. An earth-like planet must orbit its star in the habitable zone (defined in Chapter 5). The habitable zone is very narrow, and its inner and outer limits depend upon the temperature and size of the star. For cooler and dimmer stars, the habitable zone is closer in, but it is farther out for larger or hotter stars. Liquid water is a necessary component for life, so for a planet to harbor life it is necessary for it to be in the habitable zone of its star. However, there are other conditions as well. A planet must have a suitable atmosphere, and that factor depends upon the mass and size of a planet. A planet that has too little mass cannot contain any appreciable atmosphere, but a massive planet tends to hold gases that are harmful to life. Therefore, astrobiologists[11] concentrate on planets that are similar to earth in size or mass and are located in the habitable zones of their respective stars. The stars that prospective planets orbit must be of the right type too. Many stars are variable stars and are subject to frequent and intense magnetic storms, which would be deleterious to life.

---

[11] Astrobiology is a new discipline that is defined as the study of life elsewhere in the universe. There are at least two scholarly publications dedicated to astrobiology, there are astrobiology conferences every year, and there are several universities which offer degrees in astrobiology all the way up to Ph.D. Since we have no evidence of life elsewhere in the universe, astrobiology must be the only science for which there is no data.

How do the nearly 4000 extrasolar planets stack up as being possibly earth-like? We can eliminate most of them, because they do not orbit in the habitable zone. Of the remaining few, most are too large and probably have the wrong atmosphere, assuming that they have atmospheres. A very few remaining planets are in the habitable zone and may have the proper mass or size, but we cannot be sure about that. When detected via Doppler motion, we know the approximate mass of an extrasolar planet, but we do not know its size. When detected by transit observation, we know the extrasolar planet's size, but not its mass. In evaluating how earth-like a candidate planet may be, we need to know both mass and size, but we do not know both for any candidates. Assuming composition, and hence density, permits the estimation of the other quantity, but these merely are guesses. Of the very few extrasolar planets deemed to be possibly earth-like, all orbit low-MS stars, stars which have small habitable zones close to them. When orbiting this close, it is very likely that the planets in question have synchronous rotation, with one side perpetually facing the star. This would probably result in temperatures that are too hot for life on the day side of the planets and too cold for life on the night side. Only near the thin sliver of the twilight zone division between the night and day side might the temperature be conducive for life. But then there is a remaining problem. Most of the stars involved in these supposed earth-like planets are variable. Often, they are subject to chromospheric emissions that are harmful to life. The outbursts on these stars often are much greater than similar outbursts on the sun. And because these planets orbit much closer to their stars than the earth orbits the sun, the flares on these stars are far more dangerous. Many of the particle emissions would strip a planet's atmosphere. Given all these difficulties, an objective examination of the few claimed earth-like extrasolar planets reveals that we have not found any earth-like planets yet.

The data thus far strongly implies that there are no earth-like planets. One may object that not all the data are in yet, but when are the data ever all in? We always can take more data, but that normally does not deter us from making conclusions. In science, there rarely, if ever, are final conclusions. The history of science has far too many examples of ideas that were once thought beyond question but then were discarded for us to believe that almost any scientific conclusion is the final word. A sample size of nearly 4000 is large, and many conclusions are reached in science with far smaller sample sizes. Why is there such paralysis in reaching a conclusion in this matter? The ultimate aim of the search for extrasolar planets is to prove that life is common in the universe. As previously discussed, belief in evolution comes with the expectation that life is common in the universe. The proper conclusion regarding the existence of earth-like planets is contrary to the assumption of evolution, so most scientists resist this conclusion.

A few decades ago, before any extrasolar planets were found, if one were to ask most scientists how many earth-like planets we would have found by the time that we had discovered 4000 extrasolar planets, most scientists would have confidently predicted the discovery of many earth-like planets, possibly in the hundreds. This amounts to a prediction of the evolutionary worldview. On the other hand, what do creationists generally say? The Bible does not address the question of other habitable planets (or the question of extraterrestrial beings), but we can draw inferences from Scripture. For instance, in Chapter 12 of *The Created Cosmos: What the Bible Reveals about Astronomy*, I argued that there is no life elsewhere in the universe. If God has created everything, why would He have made earth-like planets when we expect there not to be any life there? This conclusion happens to agree with good science. The law of biogenesis, that living things come only from other living things, precludes the possibility that life arises spontaneously. At least two books[12] have drawn attention to the incredible characteristics of earth that make it unique.

Therefore, we have two very different predictions from the evolutionary model and the creationary model. Evolution predicts that life and earth-like planets are common in the universe. We have already seen that the data now available agrees very well with the creationary prediction but disproves the evolutionary prediction. But with regard to intelligent life, there are more lines of evidence to consider. In 1960, the astronomer Frank Drake conducted the first *SETI (Search for Extra Terrestrial Intelligence)* experiment by attempting to eavesdrop on alien radio transmission. The pace of SETI has increased sharply since then, with several SETI experiments running almost continually now. SETI probably has generated terabytes of data. Out of all that data, we

---

[12] See: P. D. Ward and D. Brownlee, *Rare Earth: Why Complex Life Is Uncommon in the Universe* (Copernicus Book: New York, New York, 2000). Also, see G. Gonzalles and J. Richards, *The Privileged Planet: How Our Place in the Cosmos Is Designed for Discovery* (Regnery Publishing: Washington D.C., 2004).

have found no evidence of any alien transmissions. This, too, is consistent with the creationary prediction, but not the evolutionary prediction.

Finally, we have the Fermi paradox, attributed to the famous physicist Enrico Fermi during an informal lunch conversation that took place around 1950. The topic of conversation was extraterrestrial, to which Fermi reportedly asked, "Where is everybody?" By this, he meant that if there were other civilizations, we would expect that some were much more advanced than our civilization, and that they long ago would have ventured into space. Therefore, we would expect that the aliens (or, at the very least, robotic craft that they built) ought to have arrived already. We see no evidence of any exploration or colonization by alien civilizations on earth or anywhere else. This, too, is confirmation of the creationary prediction, but disproves the evolutionary prediction.

There is one other pertinent point with regard to the discovery of extrasolar planets. At the end of Chapter 5, I briefly described the evolutionary theory of the origin of the planets of the solar system through the amalgamation of small bodies called planetesimals. This theory makes some predictions of a sort. That theory requires that planets that form near the sun be small and rocky, while planets that form farther away from the sun be large and gaseous. This works well within the solar system, and this has become the basis for understanding why there are two types of planets, Jovian and terrestrial. It was expected that this sort of process would play out in other planetary systems, so that massive planets would be far from their stars and small planets would be close to their stars. Extrasolar planet discoveries have provided an abundance of evidence for large planets that orbit very close to their stars. Many of these super Jupiters are far larger than any planet in the solar system, and many orbit far closer to their stars than any planets in the solar system orbit the sun. This flies in the face of the predictions of the evolutionary model of star formation. To explain this, theorists have hypothesized many scenarios where these large planets form far from their stars and then migrate inward. The observations ought to tell us that the evolutionary theories are flawed, and they indicate how different the solar system is from other planetary systems.

CHAPTER 9

Count the Stars, If You Are Able:
The Milky Way and Other Galaxies

Our Galaxy: The Milky Way

Most stars are found in galaxies, which are vast collections of billions of stars. The sun is just one star out of approximately 200 billion stars in the Milky Way Galaxy. The Milky Way is round and flat, with a bulge in its center. The diameter of our galaxy is about 100,000 light years. The central bulge of the Milky Way is 8000–10,000 light years thick. The disk of the galaxy is no more than a few thousand light years thick. The solar system is located about halfway from the center to the edge of the disk (Figure 9.1). There also is a halo component to

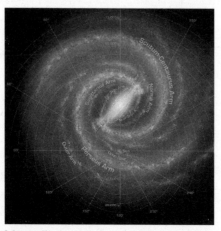

Figure 9.1. Diagram of the Milky Way Galaxy highlighting location of the solar system. (Wikimedia. Public Domain).

the Milky Way. The halo is a spherical distribution of objects centered on the galaxy. Dust and gas are common in the disk, but are nearly absent in the halo. The stars of the halo and disk are different too.

Being inside the galaxy, it is not a straightforward matter to determine the structure of the Milky Way. Serious study of galactic structure began more than two centuries ago, and we continue to refine our understanding of the Milky Way today. Studying other galaxies outside of our own helps us understand the Milky Way, because we can see the entire sweep of those galaxies.

You can see the Milky Way if you are in a dark, clear location with no moon. The Milky Way appears as a band of light stretching across the sky. Since the Milky Way is a complete circle, only a portion of the Milky Way is visible at any time. In the Northern Hemisphere, the portion of the Milky Way that is visible on summer evenings is more impressive than that which is visible during winter. Toward the southern horizon, the portion of the Milky Way visible in the sky during the summer widens. This wider portion of the Milky Way we observe is the galaxy's central bulge. The portion of the Milky Way below the horizon is not visible from temperate latitudes in the Northern Hemisphere. Similarly, there are portions of the Milky Way visible in north temperate latitudes that are not visible from south temperate latitudes. However, in much of the Southern Hemisphere the galaxy's central bulge is visible high overhead. Therefore, the view of the Milky Way from the Southern Hemisphere is far more impressive than the view from the Northern Hemisphere. Furthermore, the Large and Small Magellanic Clouds, two satellite galaxies of the Milky Way, are visible in the Southern Hemisphere. The LMC and SMC, for short, appear as two pieces of the Milky Way that have been ripped out of the Milky Way and tossed aside. Wherever your location, the Milky Way can be a grand sight in a dark, clear location. Binoculars help immensely, but most telescopes are of no help, because the field of view of a telescope usually is very small, and the Milky Way is so large.

Astronomers recognize two stellar populations (two categories of stars), population I and population II. Population I stars mostly are in the galactic disk, while population II stars are in the halo. Since the sun is in the disk, it is a population I star. In addition to location, the two populations differ in composition as well. Stars are made mostly of hydrogen and helium, with at most a few percent of everything else. Astronomers collectively call all the elements other than hydrogen and helium *metals*.[1] Population I stars have relatively more metals than population II stars, with some population I stars having 10,000 times more metals than population II stars. Population I stars often have clouds of dust and gas associated with them, while population II stars do not.

The differences in the two stellar populations are illustrated by the two types of star clusters discussed in the previous chapter. Globular clusters are comprised of population II stars, while open clusters are comprised of population I stars. The integrated colors of globular clusters are red, while open clusters are blue. This difference in color is due to what kinds of stars are brightest in either type of cluster. The light of globular star clusters is dominated by red giant stars, while blue, upper main sequence (MS) stars dominate the light of open clusters. These differences reflect the different H-R diagrams that the two types of clusters have. As discussed in the previous chapter, these differences are attributed to age, with globular clusters being older than open clusters.

Why are the compositions of the two types of clusters so different? Again, astronomers generally attribute this to age. According to the big bang model, the universe began with only hydrogen and helium, along with a very tiny amount of lithium. If this is true, then where did all the other elements come from? Recall from the previous chapter that nuclear fusion of hydrogen into helium powers MS stars. According to stellar evolution theory, once stars leave the MS, they derive their energy from other nuclear reactions. Many of those reactions in more massive stars synthesize even-numbered elements up to iron. What about the odd-numbered elements up to iron? Astronomers theorize that certain types of red giant stars (for instance, asymptotic giant branch stars) synthesize those elements. Why do these reactions terminate with iron? The nucleus of the iron atom is the most tightly bound. Consequently, nuclear fusion reactions that produce elements lighter than iron typically release energy,

---

[1] The reader will notice that this gross simplification includes some nonmetallic elements, such as carbon, nitrogen, and oxygen, among the metals. However, most elements are metals, so there is some justification in lumping everything else other than hydrogen and helium (which are not metals) together. Furthermore, the amount of metals in stars normally scales, so that the measurement of the abundance any of those elements effectively will approximate the abundance of the others. It is relatively easy to infer the abundance of some metals, such as calcium, in the spectra of stars, while determining the abundance of carbon, nitrogen, and oxygen is not.

but nuclear fusion that produces elements heavier than iron usually *requires* energy. Astronomers believe that the supernova explosions provide the energy to drive the endothermic nuclear reactions to fuse elements heavier than iron. They further think that the disruption of the envelopes of stars that accompany supernova explosions transport the products of this nucleosynthesis into the interstellar medium, the matter between stars in space. Stellar winds probably play a role in this as well.

Suppose that the universe began with only hydrogen and helium. Then the first stars to form would consist of just those two elements. As the most massive stars of the very first generation of stars aged, they would have synthesized heavier elements. Stellar winds and supernova explosions then would have spread the products of the nucleosynthesis into the interstellar medium, mixing with the hydrogen and helium gas there. Hence, the next generation of stars that formed would contain a slight amount of metals. As this process repeated, both the interstellar medium and the stars that formed from it would progressively get enriched in metals. Thus, this theory of *chemical enrichment* would serve to create all the elements that we see today. In this view, the sun and the solar system formed about nine billion years into this process, after many of the heavier elements had been synthesized. This supposedly explains the elemental abundance of the earth and even the iron in your blood, the calcium in your bones, and the carbon, nitrogen, and oxygen that make up most of your body. In speaking at a meeting of the Atheist Alliance International, Lawrence M. Krauss, author of *A Universe from Nothing: Why There Is Something Rather Than Nothing*, summed up this thinking this way:

> Every atom in your body came from a star that exploded. And, the atoms in your left hand probably came from a different star than your right hand. It really is the most poetic thing I know about physics: You are all stardust. You couldn't be here if stars hadn't exploded, because the elements—the carbon, nitrogen, oxygen, iron, all the things that matter for evolution and for life—weren't created at the beginning of time. They were created in the nuclear furnaces of stars, and the only way for them to get into your body is if those stars were kind enough to explode. So, forget Jesus. The stars died so that you could be here today.[2]

What a stark contrast this is to the biblical account of creation. The big bang model (a subject discussed in the next chapter) combined with the concept of

[2] See https://www.youtube.com/watch?v=7ImvlS8PLIo#t=16m49s, beginning at 16:50.

chemical enrichment is the ultimate evolutionary idea. Some Christians foolishly embrace these evolutionary ideas, thinking that somehow they are compatible with Scripture. However, Krauss and other atheists and agnostics understand what is going on—evolutionary ideas are an attempt to explain *everything* wholly apart from *any* Creator. Hence, these ideas are atheistic at their root.

In the 1950s, when astronomers began to realize that there were different stellar population types, they thought of them as two distinct bins. However, astronomers came to realize that population types are on a spectrum, from the most extreme population II stars (the oldest) to the most extreme population I stars (the youngest). Globular clusters are comprised of extreme population II stars, while open clusters range from old population I stars to extreme population I stars. The sun is an intermediate population I star. As low in metal content as extreme population II stars are, they do contain some metals. But the theory of cosmic evolution outlined above indicates that there ought to be stars with *no* metal content. Astronomers have dubbed these hypothetical first stars to form population III. Astronomers have searched for population III stars, but they have failed to find them. Why are there no population III stars? One suggestion is that all the population III stars were massive, and since massive stars have short lifetimes, no population III stars exist today. Another suggestion is that supernova explosions and stellar winds transported processed material to the photospheres of any remaining primordial stars, effectively contaminating them with some metals, so that there are no metal-free stars today. Or perhaps God created the spectrum of population types for His reasons, but man, in his naturalistic thinking, has concocted a false cosmic history.

According to this naturalistic history, how did the galaxy form? For a long time, astronomers have thought that the galaxy formed from a large spinning cloud of gas. As the cloud contracted, the gas fragmented into pieces that formed into stars. Eventually, the gas flattened into a disk. Stars that formed early in this process would preserve their orbits in the galactic halo, but stars that formed later would orbit in the disk of the galaxy. The first stars to form would have few metals, while stars that formed later would have more metals. This would explain the location and metallicity of stellar populations. Dust could not form until a sufficient amount of heavier elements formed,[3] which is

---

[3] Dust generally does not contain helium or hydrogen, though some hydrogen exists in dust as part of ice. However, ice also requires that oxygen exist, which, according to the theory, was created later.

why dust is found in the disk, but not the halo. This view of galaxy formation is referred to as top-down, because it begins with a large structure, the large gas cloud, that formed the smaller structures that we see in the galaxy today. However, in recent years, some astronomers have proposed that galaxies form in a bottom-up manner. That is, stars began to form, after which they began to amalgamate into larger groupings that eventually formed into galaxies.

Extragalactic Astronomy

Extragalactic astronomy is the study of objects outside of the Milky Way Galaxy. Most of these other objects are galaxies. How many galaxies are there? Until recently, astronomers had thought for several decades that the universe contained at least 100 billion galaxies. In its simplest form, this estimate was based upon counting the galaxies that our most sensitive telescopes and instruments could detect in a tiny patch of the sky and dividing that number by the fraction of the entire sky that the patch represented. However, in October 2016, astronomers concluded that we had undersampled the number of galaxies by at least a factor of ten. Based on this new work, there probably are at least two trillion galaxies.

Galaxies come in various sizes. The Milky Way is a typical large galaxy. However, most galaxies are much smaller *dwarf galaxies*. A few very rare galaxies are exceedingly large, much larger than galaxies like the Milky Way. Galaxies are not uniformly distributed in space; rather, most galaxies congregate into groups called galaxy clusters. The Milky Way is a bit unusual in that it is not directly part of a galaxy cluster. Instead, it is a part of what astronomers call the *Local Group*. The Local Group consists of three large galaxies, the Milky Way, the Andromeda Galaxy (also called Messier 31, or M31),[4] and the Triangulum Galaxy (M33), as well as several dozen dwarf galaxies. The closest galaxy cluster is the Virgo Cluster, a collection of more than a thousand galaxies centered about 55 million light years away. The Milky Way, along with several other groups and small clusters, are included with the Virgo Cluster into the Virgo Supercluster. The Virgo Supercluster is about 100 million light years across, with the Virgo Cluster at its center. There are millions of superclusters

---

[4] A little more than a hundred of the brighter appearing nebulae, star clusters, and galaxies are in the Messier catalog, compiled by the French astronomer Charles Messier in the late eighteenth century. A Messier object usually is referred to by the capital letter M, followed by its number in the catalogue.

of galaxies in the universe. The hierarchical structure continues upward, with the Virgo Supercluster being a part of an even larger Laniakea Supercluster (it is about 500 million light years across). I shall return to this hierarchical structure of the universe later in this chapter.

When comparing photographs of galaxies, it is very clear that there are two basic types of galaxies—spirals and ellipticals (Images L, M, N, O, P, and Q). As the name suggests, elliptical galaxies have elliptical shapes, ranging from circles to very flattened ellipses. There is a central condensation of stars, with stellar density decreasing with increasing distance from the center. The Milky Way is a spiral galaxy. Like the Milky Way, spiral galaxies have disks with central bulges surrounded by halos. The disks of spiral galaxies have spiral arms that gracefully curl outward from their nuclei. In some spiral galaxies, the spiral arms come off the end of a bar that passes across their nuclei. Astronomers call these barred spiral galaxies. Because dust in the galactic plane between us and the galactic center absorbs light, it is difficult to probe the center of the Milky Way. However, the data indicate that the Milky Way is a barred spiral galaxy.

In the 1920s and 1930s, Edwin Hubble did pioneering work in extragalactic astronomy. He noted the two basic types of galaxies, and he further classified galaxies according to their morphology. For instance, Hubble noticed that spiral galaxies, both barred and normal spirals, varied in how tight their spiral arms were wound, and that the size of their nuclei likewise varied. More importantly, he realized that galaxies with larger nuclei tended to have tighter spiral arms, so Hubble made this correlation the basis of his classification of spirals as a, b, or c. Therefore, a Hubble type Sa galaxy has a large nucleus with tight spiral arms; an Sc galaxy has a small nucleus with loose arms; while an Sb galaxy is intermediately between the two. For the barred spiral galaxies, Hubble created the types SBa, SBb, and SBc parallel to the types for normal spiral galaxies. Hubble classified elliptical galaxies with a capital letter E followed by a number between 0 and 7 indicating how elliptical the galaxy was, with an E0 galaxy appearing circular and an E7 galaxy being the most elliptical. There were a few galaxies, perhaps 1% or 2%, that defied classification. Hubble classified these as Irr, for irregular. For instance, both the LMC and SMC are irregulars.

Hubble arranged his galaxy types on a diagram in the shape of a tuning fork (Figure 9.2). E0 galaxies are on the far left, near the end of the handle of the tuning fork. Moving rightward toward the tines are each elliptical type up

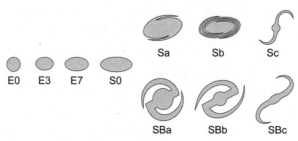

Figure 9.2. Illustration of classification of galaxies.

through E7. Beyond that are the two tines, with, in order, types Sa, Sb, and Sc on one tine, and SBa, SBc, and SBc on the other tine. Sometimes the irregular galaxies are placed to the far right. Hubble referred to galaxies to the left of the tuning fork diagram as *early* and those galaxies on the right as *late*. Apparently, Hubble meant this in terms of reading left-to-right on the diagram and not in some evolutionary sense. However, some other astronomers did interpret the tuning fork diagram as an evolutionary one, but while some thought that the evolution was from left to right, others thought that the evolution was from right to left. Notably, for at least the past half-century, astronomers have thought that neither view is correct, opting instead for galaxies generally remaining as the type that they began as. There are some notable exceptions to this general rule though. Some of the largest galaxies known are giant ellipticals, and astronomers now think that they may be the result of mergers of many galaxies.

One of the issues that helped fuel the debate as to whether galaxies evolve from one type to another is the persistence of spiral arms in spiral galaxies. The orbital motions of objects in galaxies are more complicated than first appears. Presumably, each object follows an orbit governed by gravity; so, obviously, the disk of a galaxy does not spin as a single unit. But neither is galactic orbital motion as simple as it is in the solar system. In the solar system, nearly all the mass is concentrated at the center, in the sun. This result is that planets follow Keplerian motion, with orbital speed decreasing with increasing distance. But the mass of a galaxy is not concentrated at its center. Instead, the mass is spread out in the disk and in the halo. It can be shown using calculus that if mass is reasonably symmetrically distributed, then an orbiting body is affected only by the mass lying closer to the center of motion. Astronomers take advantage of this fact and use it to probe the distribution of matter within a galaxy, a topic that I shall take up shortly.

For now, it is important to note that objects in the central region of a spiral galaxy orbit with speeds that are directly proportional to their distance from the center. The result is that the nuclear regions of spiral galaxies spin nearly as a unit. However, in much of the disk, and especially where the spiral arms are, the orbital periods of objects decrease with increasing distance. Therefore, the outer regions of spiral arms ought to orbit less often than the inner regions. After a few rotations, the spiral arms ought to be smeared out so as to be unrecognizable. Given the orbital speed of the sun (about 150 mi [250 km] per second) and its orbital radius (approximately 25,000 light years), the orbital period of the sun is roughly a quarter billion years. Using this as a good average rotation speed, the Milky Way would experience eight rotations in just two billion years. Astronomers think that the Milky Way and other spiral galaxies are more than 10 billion years old, so why do these galaxies still have spiral arms? Of course, biblical creationists, who believe that the universe is only thousands of years old, see the spiral galaxy wind-up problem as evidence that the universe is very young.

As you might imagine, evolutionary astronomers are aware of this problem, so they have proposed solutions to it. But first, I need to address what spiral arms are and what they are not. When looking at a photograph of a spiral galaxy, it is easy to think that the spiral arms represent the locations of much of the matter in the disks of galaxies, but this is not the case. If one were to conduct a complete stellar census and plot the locations of all the stars in a spiral galaxy's disk, the spiral arms would not show up. What causes spiral arms to show up is that the very hottest and brightest stars (spectral types O and B) are congregated along the arms (Image R). While these stars represent only a tiny fraction of the total mass of stars, they dominate in the light emitted by stars. The regions between spiral arms lack these bright beacons, but the density of stars is about the same between the spiral arms as it is along the arms. Astronomers call objects that correlate, and hence mark, the locations of the spiral arms *spiral tracers*. Although not as obvious as the O and B stars, other spiral tracers include various types of gas and dust clouds. This is because the density of the interstellar medium is much greater in the spiral arms than between the spiral arms. Astronomers think that the large, cool gas clouds common in the spiral arms are the sites of star formation, and thus stars are born in the spiral arms. As we saw in the previous chapter, O and B stars have very short MS lifetimes, so they cannot travel far from the places of their births. Therefore, astronomers reason, it is no accident that O and B stars are found along spiral arms.

Why are there so many gas clouds in the spiral arms? In the 1960s, astronomers developed density wave theory to explain this, and also to explain the persistence of spiral arms. According to this theory, there is a density enhancement in the galactic disk. The complex gravitational field of the galaxy interacts with the density enhancement to give the density enhancement its characteristic spiral shape. The result is supposed to be a sort of standing wave in the disk of a spiral galaxy. Most objects that orbit in the disk enter and leave the density wave with little effect. The exceptions are clouds of gas and dust, which are compressed. This compression supposedly produces the clouds found along the spiral arms. The compression might also be responsible for some star formation that supposedly is common along the spiral arms. It is important to note that the spiral arms themselves move very little while other material in the disk does. Eventually, astronomers began to invoke the influence of the gravity of satellite galaxies to maintain the density wave. Now it is fashionable to appeal to dark matter as playing a significant role. The invocation of these additional factors amounts to a tacit admission that density wave theory probably does not adequately explain spiral structure.

One of the first things that astronomers want to determine about other galaxies is their distances. The LMC and SMC are approximately 160,000–180,000 light years away, but they merely are satellites of the Milky Way. The closest galaxy outside of ours and comparable in size to the Milky Way is M31. It is a little more than two million light years away. How do astronomers measure these distances? This distance is much too far for trigonometric parallax, so astronomers must employ indirect methods. In the previous chapter, I discussed one of these, the Cepheid variable method. Cepheid variables are giant and supergiant stars, so they are very bright and hence stand out in other galaxies. We can use the Cepheid method out to about 50 million light years. In addition, astronomers have developed *standard candles* to determine extragalactic distances. A standard candle is an object for which we think we know its absolute magnitude. If we can identify a standard candle in another galaxy and measure its apparent magnitude, then we can measure its distance. Some of the standard candles are:

- The brightest supergiant stars
- Bright novae
- Brightest globular star clusters
- Brightest H II regions
- Type Ia supernovae

An H II region is a bright cloud of ionized hydrogen.[5] Astronomers can use both the brightness and the apparent sizes of the brightest globular clusters and brightest H II regions to measure the distances of their host galaxies. Type Ia supernovae are a class of supernovae. Astronomers have both observational and theoretical reasons for thinking that type Ia supernovae have the same absolute brightness at maximum light. When they can, astronomers will use several methods of measuring the distance to a galaxy, and then average the results.

Probably the best-known method of finding the distances to galaxies is by measuring redshift and using the Hubble relation. This relation relies upon the expansion of the universe, discovered by Edwin Hubble in 1929. To illustrate, consider a source of sound, such as a siren on an emergency vehicle. If the siren moves toward us, its sound waves are compressed, and we hear a higher pitch. However, if the siren moves away from us, we hear a lower pitch, because the sound waves are stretched.[6] This is called the Doppler effect. A similar thing happens with moving sources of light, such as stars. Motion toward us shifts the star's spectrum toward shorter wavelengths. In visible light, shorter wavelengths are blue, so we call this a blueshift. Conversely, if a star moves away from us, its light is shifted toward longer wavelengths, toward the red end of the visible spectrum, so we call this a redshift. Astronomers measure blueshift or redshift of stars by how much the absorption lines in their spectra are shifted from the wavelengths they normally would have if the stars were not

---

[5] The *H* in H II refers to hydrogen, while the Roman numeral *II* refers to the hydrogen being singly ionized. The Roman numeral *I* refers to un-ionized hydrogen. Astronomers refer to all ionized atoms this way, with larger Roman numerals indicating successively higher levels of ionization. Since hydrogen has only one electron, there is no H III. H I regions also exist, but we generally detect them in the radio spectrum. H I regions are found in the spiral arms, and hence are spiral tracers. They have played an important role in deducing galactic structure.

[6] A common misconception is that the Doppler effect affects loudness. However, the Doppler effect only alters pitch, though eventually the changing distance would change the loudness too.

moving. When measuring the motion of stars within our galaxy, about half are redshifted, while half are blueshifted (Figure 9.3). However, nearly all galaxies are redshifted. The only exceptions are a few galaxies that are very close to us.

If the universe is expanding, then we would expect most galaxies to have redshifted spectra.[7] Furthermore, there ought to be a linear relationship between the amount of redshift and distance. We can write this mathematically as

$$V = HD$$

where V is the amount of redshift and D is the distance of a galaxy, while H, called the Hubble constant, is the constant of proportionality. Hubble measured the redshifts and distances of several galaxies and found that there was a linear relationship. A plot of the data is the best way to illustrate this linear relationship, with H being the slope of the line that best represents the data (Figure 9.4). Hubble's first estimate of the Hubble constant was 550 km/s/Mpc.[8] While measurement of redshift would appear to be straightforward, measurement of extragalactic distance is not. Hubble soon reduced the value of the Hubble constant, as did other astronomers. By 1960, the agreed upon value of the Hubble constant was about 50 km/s/Mpc, where it remained for three decades. However, in the 1990s, estimates of H increased for the first time. Today, most astronomers think that the value of H is about 70 km/s/Mpc (13 mi/sec for every million light years), though the exact value is still debated.

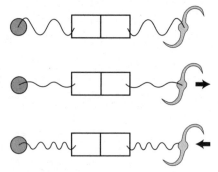

Figure 9.3. Diagrams of simulated spectra (normal, red shifted, blue shifted).

[7] Technically, redshift due to universal expansion is not due to the Doppler effect, but is observationally indistinguishable from a Doppler redshift.

[8] Spelled out, these units are kilometers/second/megaparsec. A megaparsec is a million parsecs, about 3.26 million light years.

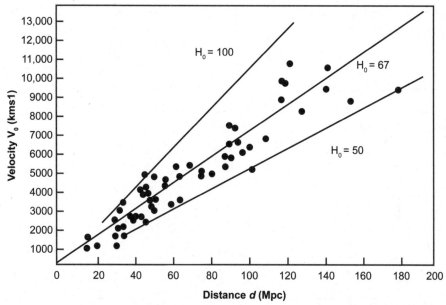

Figure 9.4. Data plot expressing linear relationship of the Hubble constant.

Why has there been so much disagreement about the value of H? Historically, much of the disagreement was over the accuracy of distance measurements, but there is another subtle uncertainty regarding redshifts that has become more significant in recent decades as uncertainty concerning distances has decreased. If galaxies were motionless with respect to space, there still would be the Hubble law, because universal, or cosmic, expansion is due to expansion of space itself. That is, even if galaxies were not moving, they would get farther apart because the space between them is increasing in size. However, galaxies are moving through space. This complicates matters, because the measured redshift of a galaxy is a combination of cosmic expansion and Doppler shift due to true motion of the galaxy with respect to space. While cosmic expansion is always a redshift, the Doppler motion is a redshift half the time and is a blueshift half the time. Since cosmic expansion increases with increasing distance, nearby galaxies have very feeble redshifts, while distant galaxies have much greater redshifts. This is why a few nearby galaxies have observed blueshifts rather than redshifts—the Doppler motion happens to be toward us, and it exceeds the amount of cosmic redshift. To determine H, we need to use only the cosmic expansion portion of redshift. However, we do not know how much of the observed redshift of a galaxy is due to cosmic expansion and how much is due to the galaxy's Doppler motion. The normal way to avoid this difficulty is to use the redshift of distant galaxies, because while cosmic redshift increases with distance, Doppler motions do not, but instead are random about some maximum value. However, while the redshift of distant galaxies is more likely to be dominated by cosmic expansion, there is more uncertainty in their distances. Much of the debate for decades has concerned how to handle this conundrum.

Why is the knowledge of the Hubble constant important? There are two reasons. One reason is that it measures the rate of cosmic expansion. That rate is significant in developing cosmological models, the topic of the next chapter. The second reason is that we can turn the Hubble relation around to find distances to galaxies. Rearranging the Hubble law, we get

$$D = V/H.$$

If we know the value of H, then all we must do to find the distance of a galaxy is to measure its redshift. A common misconception about the Hubble relation is that it can be used to measure distances of all sorts of objects, such as stars. However, the Hubble relation only works for extragalactic objects, such as galaxies, and even then it is reliable only for large distances.

Some biblical creationists express doubt about the Hubble relation. There probably are two reasons for this. First, the Hubble relation is used to measure the distance of most galaxies, because most galaxies are too far away for the other methods to work, and type Ia supernovae are such rare events that they have not been observed in most galaxies. Consequently, the Hubble relation generally gives the largest distances that we encounter in astronomy. These extremely large distances are what prompt most people to ask about the light travel time problem.[9] In the minds of some people, casting doubt upon distance determination methods in astronomy might be a good response to the light travel time problem, but it really is not. The second reason many biblical creationists may doubt the Hubble relation is its perceived connection to the big bang model (which I shall discuss in the next chapter). The Hubble relation is a consequence of an expanding universe, and the big bang model is

---

[9] I will not discuss the light travel time problem in this book. For a discussion of that topic, please see Chapter 11 of *The Created Cosmos: What the Bible Reveals about Astronomy* (Danny R. Faulkner. Green Forest, Arkansas: Master Books, 2016).

dependent upon the universe expanding. Some biblical creationists may think that the denial of an expanding universe is a sort of silver bullet argument against the big bang. Indeed, if the universe is not expanding, then the big bang model is not viable. However, what if the universe is expanding? The big bang is only one possible cosmogony, for there are many other possibilities, including some biblically based ones. It would be a shame to throw out an important clue in developing a good biblical model. The Hubble relation is an observational fact. The most straightforward interpretation is that the universe is expanding. It behooves anyone who doubts that interpretation to provide an alternate explanation for the Hubble relation.

Once we know the distance of a galaxy, we can determine other properties. For instance, we can measure the total light of a galaxy and express it as an apparent magnitude. If we know the distance to a galaxy, then we can convert the apparent magnitude to absolute magnitude. Absolute magnitude generally is related to mass, because more mass usually translates into more stars, which results in more light.

Dark Matter

Measuring masses of galaxies is difficult. As with stars, we measure galactic mass by observing orbital motion. In the case of galaxies orbiting one another, such as the two companion galaxies to M31, the orbital motion can give us some information about the mass of M31. However, this information is limited, because the orbital plane is unknown. More helpful are rotation curves of galaxies. A rotation curve is a plot of the Doppler motion of objects within a galaxy measured at various distances from the center of the galaxy. Objects on one side of the center are redshifted (treated as positive values), while objects on the other side are blueshifted (treated as negative values). This data can be inverted and combined to produce a single plot of positive data (Figure 9.5). The radial velocity increases linearly with increasing distance from the center at first, before reaching a peak. Beyond the peak, astronomers expected the rotation curves to decrease at greater distances. By using values near the peak (velocity and distance from the center), astronomers could estimate the mass of a galaxy.

In the mid-1970s, astronomers began to investigate the orbital behavior of objects beyond the peaks of rotation curves of galaxies. To their astonishment,

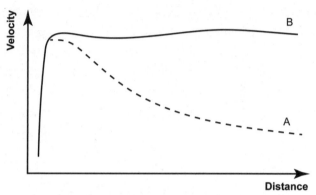

Figure 9.5. Data plot of rotation curve. (Wikimedia).

they found that not only did the orbital speeds not decrease as expected, the speeds increased in some cases. This trend continued out to the visible edges of the galaxies and beyond where nearly all the light of the galaxies came from. This is a strange result, and there has since been much debate as to what this means. There have been several suggested explanations, and all of them represent radical departures from known physics. For instance, one suggestion is modified Newtonian dynamics (usually abbreviated as MOND). MOND proposes a change to the inverse square law of gravity at very great distances. Most astronomers consider *dark matter* to be far less radical. Dark matter is a hypothetical substance that interacts with normal matter only via gravity. Most striking is that dark matter does not emit radiation as normal matter does, hence the name. Dark matter is invoked to explain the anomalous extra gravity that galaxy rotation curves indicate. If this theory is correct, then most of the mass of galaxies consists of this mysterious dark matter. At most, normal matter may comprise 20% of the mass in galaxies.

Another line of evidence for dark matter goes back to the 1930s. Fritz Zwicky measured the Doppler motions of galaxies within the Coma Cluster (another galaxy cluster more than 300 million light years away). From the dispersion in the Doppler motion of its members, Zwicky measured the mass of the Coma Cluster. This is called the *dynamic mass*. Zwicky also measured the *lighted mass*, from the total light emitted by the Coma Cluster's members. As previously mentioned, more mass translates into more stars and hence more light. Astronomers had worked out a mass-to-light ratio from studies within our own galaxy, so dividing the light of the Coma Cluster revealed the lighted mass. Zwicky found that the dynamic mass of the Coma Cluster exceeded the lighted mass by a factor of ten. He and other astronomers found similar results when they studied other galaxy clusters. For a half century, this *missing mass problem* was viewed as an oddity, but with a second line of evidence from rotation curves of galaxies obtained in the 1970s, astronomers began to realize that this was more than just a minor problem.

There are still other lines of evidence for dark matter's existence. Nearly a century ago, physicists realized that Einstein's theory of general relativity indicated that massive objects can form images of more distant objects, an effect called *gravitational lensing*. The first image from gravitational lensing was discovered in 1979, and many more have been found since. Some examples of gravitational lensing are arc-like images of distant galaxies imaged by galaxy clusters. By studying the multiple arced images of a galaxy, astronomers can compute the mass required to produce the images. Those measurements match the amount of additional mass indicated by other means.

What is dark matter? We do not know. Many candidates have been proposed, but most of them have been eliminated. Dark matter may consist of an entirely new type of particle that we do not yet know about. Particle physicists do not have room for a new particle in their models, so if dark matter is real, it could cause a revolution in physics. Indeed, the person who solves this mystery probably will receive a Nobel Prize in physics.

Some biblical creationists express doubts about dark matter. They appear to be motivated by the manner in which dark matter has been invoked in the big bang model. Theorists have used dark matter to solve conundrums in cosmology, such as questions about how galaxies form, ideas of how spiral arms persist, and a parameter that helps make the big bang work better. It is easy to think that dark matter has been invented just to solve these problems, and this may be the reason why many biblical creationists prefer MOND over dark matter. However, astronomers resisted the evidence for dark matter for many years. It was only *after* astronomers realized the reality of dark matter that they embraced it and then used dark matter to solve other problems. There are good reasons for believing that dark matter is real. What may motivate some biblical creationists in rejecting dark matter is concern over losing an argument for recent creation. If there is no dark matter, then the galaxies within the clusters are not gravitationally bound, and galaxy clusters will disperse in a few hundred million years. This would indicate that clusters of galaxies, as well as the galaxies they contain, are far younger than billions of years. However, our motivation ought not to be maintaining our model.

Active Galactic Nuclei

An *active galactic nucleus* (AGN) is a region in the center of a galaxy that produces an unusually high luminosity. An *active galaxy* is the galaxy that hosts an AGN in its core. AGNs can be extraordinarily bright in any part of the spectrum, from radio waves to gamma rays. Nor are these portions of the spectrum exclusive, as some AGNs are active over a wide range of wavelengths, with a few possibly encompassing the entire spectrum. Astronomers have classified many different types of AGNs. Many of the earliest recognized AGNs were grossly over-luminous in the radio part of the spectrum (referred to as radio-loud), so radio surveys of the sky served to identify many AGNs easily. However, astronomers eventually found AGNs that were radio-quiet. The distinction between radio-loud and radio-quiet remains a useful means of classification of AGNs. The first radio-quiet AGNs discovered were *Seyfert galaxies*, named for Carl Seyfert, the astronomer who first recognized the class in 1943. Seyfert galaxies are spiral galaxies that appear normal, but have very bright, yet very tiny, nuclei. Seyfert galaxies are one of the two largest groups of active galaxies.

The other large groups of active galaxies are *quasi-stellar objects*, often abbreviated QSOs, but frequently known as the portmanteau *quasars*. Many of the early quasars were radio-loud, but eventually astronomers found many quasars that are radio-quiet. During the early days of radio astronomy, much of the work was surveys of the sky, with attempts to identify optical counterparts to radio sources discovered. In many cases, the optical counterparts were galaxies and some nebulae. A few optical counterparts were faint blue stars. Stars do not produce significant radio emission, so this was strange, and it led to these objects briefly being called *radio stars*. In 1961, a spectrum was taken of 3C 273,[10] the brightest of the radio stars. The spectrum was largely featureless, except for a few bright emission lines. At the time, no one knew what to make of the spectrum. It was not until two years later that an astronomer realized that these

[10] The name indicates that this object was entry number 273 in the *Third Cambridge Catalogue of Radio Sources*, a survey of radio sources published in 1959.

emission lines were due to hydrogen but were not identified earlier because the lines were greatly redshifted from their normal wavelengths. The redshift was nearly 16%, at the time among the greatest redshifts known. Assuming that this redshift is cosmological,[11] the Hubble law results in a distance of about 2.5 billion light years. Using this distance and the measured apparent magnitude, the absolute magnitude is –26.7. This is about five magnitudes brighter than the absolute magnitude of a galaxy like the Milky Way. Recall that a difference of five magnitudes corresponds to a factor 100 in brightness, so 3C 273 is about 100 times brighter than the entire Milky Way, a collection of a few hundred billion stars. Therefore, 3C 273 outshines more than a few tens of trillions of stars.

This luminosity is typical of quasars. Yet quasars appear very small. Indeed, they are far, far smaller than galaxies. Quasars and other AGNs vary in brightness. Indeed, prior to the discovery of the faint glow around them, some AGNs were mistaken for variable stars before their true identity was discovered.[12] Variability must have some physical mechanism, which requires that it can operate only so fast. The maximum speed for physical processes is the speed of light. Hence, the period of variability ought to be the maximum size, expressed as light travel time. Some quasars vary in brightness over a few years or even months, so their maximum size is a few light years or a few light months. Keep in mind that this is the maximum size—the actual size probably is much smaller. This presents us with a problem. Quasars produce more energy than trillions of stars, yet they are only slightly larger than the solar system. What can their source of energy be?

There is only one possible engine that we know of that can meet the requirements—a supermassive black hole that is accreting matter at a large rate. What is meant by *supermassive*? Supermassive black holes typically contain *at least* a million times the mass of the sun, but some contain a billion solar masses. As the matter falls inward, part of the matter is sacrificed to the black hole, in exchange for energy that produces the extreme brightness in such a small volume. Due to angular momentum considerations, we would expect

---

[11] A redshift being cosmological means that the redshift is due to expansion of the universe, and hence the Hubble law applies in finding the distance. The distance quoted here is based on the latest value of the Hubble constant.

[12] For instance, BL Lac objects are a class of AGNs. They are named for the prototype of the class, BL Lacertae, which was the first member of the class studied in some detail. The name BL Lacertae is a variable star designation.

that the in-falling matter would flatten to an accretion disk. Highly energetic magnetic effects would result in jets of matter and radiation from the center perpendicular to the plane of the accretion disk. These accretion disks and jets are similar to those found in X-ray binaries discussed in the previous chapter, though on a much grander scale.

In the early history of the study of AGNs, there appeared to be a bewildering array of different objects, each with their own peculiar traits, such as overall luminosity, variability, which parts of the spectrum were involved, the presence of jets, and what kinds of spectral lines were present in the spectra. However, there slowly developed a more holistic theory that might explain all AGNs. Different factors could explain the differences in what we observe in different types of AGNs. For instance, since an AGN's luminosity is how much energy is radiated, the rate at which matter falls into the black hole must be the deciding factor for luminosity. Therefore, the brighter AGNs must have greater mass in-fall rates. If the accretion disk is very thick, it would prevent jets from spreading, so a thick jet may act to focus a jet into a tight beam. It also might prevent us from seeing down toward the central engine. On the other hand, a thin disk might permit a broader beam. The orientation at which we view an AGN would alter what we see. The view looking face-on to the disk would be very different than the view from the side. Physical factors might intensify the presence of jets, or they may mute jets. Sometimes only one jet may be visible rather than two. Many of these factors change with time, so differences between different types of AGNs may be due to age. For instance, at 2.5 billion light years, 3C 273 is one the closest quasars, making quasars sparse in our region of space. However, the density of quasars is much higher at greater distances. If the universe is billions of years old, then this is attributed to the aging process of quasars. Astronomers believe that quasars are the cores of very young galaxies. While galactic nuclei are very active in their youth, they settle down with age. At great distance, we see many young galaxies, but in our vicinity of the universe, we see only mature galaxies, so quasars do not exist locally.

There is some evidence for this scenario, besides the overlapping properties of the different types of AGNs. Back in the 1960s, some astronomers suggested that quasars were the active nuclei of galaxies, though there was no evidence of that at the time. One of the predictions of this proposal was that much deeper photographs would reveal that quasars are surrounded by much fainter light emanating from the rest of the host galaxies. Later observations proved this to be correct. Many nearby galaxies, including the Milky Way, appear to be normal, at least from the standpoint of the activity of their nuclei. However, in recent years, astronomers have been able to measure the motions of things orbiting very close to the centers of some of these galaxies. The observed motions require huge mass to be enclosed in the very tight confines of the galactic cores. In the case of the Milky Way, the mass required to explain the orbital motion is 4 million solar masses. For the case of M104, a galaxy about 30 million light years away and part of the Virgo Cluster, the required mass is about a billion solar masses. There is nothing visible at the centers, and the volume is too small to contain any normal objects of this mass anyway. Therefore, astronomers conclude that these "normal" galaxies have supermassive black holes lurking in their cores. In fact, it appears that all large galaxies have supermassive black holes in their cores, though they may not be active. In their infancy, the cores typically are very active, with much matter being fed into the black holes. However, with time, apparently nearly all the matter readily available to the black holes is consumed, and they have a relatively quiescent existence in their maturity.

If this scenario sounds evolutionary, it is. What alternate scenarios have creationists developed? Unfortunately, not much, because very little has been written in the creation literature about extragalactic astronomy and AGNs. This lack of a good response may be part of the motivation to reject the existence of black holes discussed in the previous chapter. However, the ramifications of the proposed origin and evolution of black holes that the explanation of AGNs seems to demand ought not to negate the good operational science that strongly suggests that black holes, even supermassive ones, exist.

Also related to this may be the embrace of the work of the late astronomer Halton Arp. When quasars were first discovered in the 1960s, Arp was alarmed by the rapid rush to accept the great distances of quasars and the subsequent need for an exotic source for their power. Arp proposed that not all redshifts are cosmological, and so it followed that quasars were much closer than generally thought. Over the years, Arp produced many observations calling into question the automatic interpretation that large redshift implied great distance. Arp argued that there may be a significant non-cosmological component to some redshifts. Many biblical creationists uncritically accepted much of what Arp said in a misguided attempt to discredit the big bang.[13] Contrary to the popular misconception, Arp did not doubt the expansion of the universe, or that, in general, redshifts were related to distance for galaxies. He simply doubted that quasar redshifts were cosmological. Even among astronomers, Arp's views remain controversial. While I find some of Arp's work interesting, I believe that among biblical creationists, his contribution is overrated. There are difficulties with his work that many biblical creationists ignore.

I wish that I had more positive things to say about the ideas of AGNs at this time, but I do not. However, much of this is due to lack of work in the field, so there is reason for optimism. Hopefully, with young creationists entering the field of astronomy, some can prepare themselves in the area of extragalactic astronomy to better address these issues.

---

[13] This is an excellent example of how worldview affects the interpretation of evidence.

CHAPTER 10

A Contemporary Perspective:
Modern Cosmology and the Bible

Cosmology is the study of the structure of the universe.[1] A related term is *cosmogony*. Cosmogony is the study of the origin and history of the universe. Much of what is called cosmology today actually is cosmogony, but that is a fine point that is largely lost. Hubble's discovery of the expansion of the universe in 1929 is the cornerstone of modern cosmology. Contrary to common belief, Hubble did not stumble upon this discovery. Rather, he was aware of some important developments that caused him to anticipate the result. In 1912, the American astronomer Vesto Slipher discovered that most galaxies had redshifts, though no one at the time understood what that meant. For one thing, as we saw in Chapter 7, most astronomers then thought that galaxies were clouds of gas within the Milky Way. But in 1924, Hubble demonstrated otherwise. About the same time, it was shown that Einstein's theory of general relativity, when applied to the universe, results in a universe that is either expanding or contracting. Slipher's work seemed to confirm that the universe was expanding. Being aware of this, Hubble realized that the next logical step was to see if redshift and distances of galaxies were linearly related, as required by the expanding universe.

---

[1] This chapter is only a brief discussion of cosmology. If you would like a more involved treatment of cosmology within a biblical creation framework, please see *Universe By Design: An Explanation of Cosmology and Creation* (Danny R. Faulkner, Green Forest, Arkansas: Master Books, 2004).

Once the matter of the expanding universe was established, astronomers naturally wondered about the past. If the universe is expanding, then its overall density is decreasing. However, since the time of the ancient Greeks, scientists in the West generally thought that the universe was eternal. If the universe were expanding and eternal, then the density of the universe ought to be zero, but clearly it is not. To salvage the eternal universe, by the mid-twentieth century cosmologists had developed the steady state theory. The steady state theory posited that density is conserved as the universe expands. A constant density in an expanding universe required that new matter spontaneously appear. Because matter is constantly created, the steady state theory sometimes was called the continuous creation model. One might object that the continuous creation of matter would violate the conservation of matter, but proponents of the steady state model countered that they wished to replace the conservation of matter with the conservation of matter density in the universe. The required rate at which new matter was introduced to maintain constant density per unit volume was extremely low, rendering it difficult to determine observationally which principle was true.

Not only did many scientists think that the universe was eternal, but many scientists also thought that the universe was infinite as well. However, this presented a problem. If the universe were eternal and infinite, and if stars were reasonably uniformly distributed throughout space, then in every direction that we looked, our view eventually would be blocked by the surface of a star. The angular surface areas of stars decrease with the square of the distance. At the same time, the number of stars visible increases with the square of the distance. These two effects exactly cancel. The result is that in an eternal, infinite universe where stars are reasonably uniformly distributed, the entire sky ought to be as bright as any star, such as the sun. This is starkly at odds with how dark the night sky is. This observation is known as *Olber's paradox*, for Heinrich Wilhelm Olbers, a German astronomer who discussed this topic nearly two centuries ago, though others had discussed it in the previous two centuries. For a long time, Olber's paradox was a problem, and many different solutions were proposed. None of them worked. Ultimately, at least one of the assumptions upon which Olber's paradox was based—that the universe is eternal—turned out to be wrong. Some biblical creationists still bring up Olber's paradox supposedly as a problem for modern cosmology, but it is not relevant anymore because most astronomers now think that the universe had a beginning in the finite past. Therefore, biblical creationists ought not to use Olber's paradox as an argument.

The Big Bang Model

Why was there a change in thinking on the age of the universe, abandoning an eternal universe in favor of a universe that had a beginning in the finite past? While many astronomers and cosmologists embraced the steady state theory, others were alarmed by the jettisoning of the conservation of matter. As previously mentioned, if matter were conserved, then the density of the universe would have been greater in the past. Greater temperature would have accompanied greater density as well. These conditions could not have extended indefinitely into the past, so this reasoning led to the conclusion that the universe began in a very hot, dense state, from which it expanded and cooled into the universe that we see today. By the mid-twentieth century, this theory became known as the big bang model, though that is not a very good name, because the big bang model, if properly understood, is not an explosion. Many biblical creationists have argued against the big bang model based on the assumption that the big bang is an explosion. Improper formulation of an idea, followed by a refutation of that improper formulation of the idea is known as a straw man argument. A straw man argument is in informal logical fallacy. Biblical creationists ought to criticize the big bang model, but they ought not to use a straw man argument to do it.

These two major cosmologies, the steady state and big bang, were the major competing cosmologies by the mid-twentieth century. For a while, the steady state was the preferred model. However, this all changed in 1965 with the discovery of the *cosmic microwave background* (CMB). According to the big bang model, stars did not exist until millions of years after the universe began. All that existed at first was gas, mostly comprised of hydrogen. The gas was so hot that all the hydrogen was ionized. Light cannot travel very far in an ionized gas, so light was continually absorbed and reemitted in the early universe, rendering the universe opaque. However, approximately 380,000 years after the big bang, the universe had cooled enough that neutral atoms of hydrogen could exist for the first time. With light no longer blocked by ionized gas, this transition rendered the universe transparent for the first time. Physicists say that matter and light were decoupled, freeing light to travel immense distances. After traveling billions of years, the expansion of the universe would have greatly redshifted the light from the era of decoupling.

The light emitted at the time of decoupling would have a characteristic spectrum called a *blackbody spectrum* (Figure 10.1). The physics of blackbody spectra is well understood. A blackbody spectrum has a peak, and the wavelength of that peak is inversely proportional to the temperature. Physicists and astronomers frequently use this fact to determine the temperature of objects. An infrared thermometer makes use of this as well. The extreme redshift that the radiation from the age of decoupling experienced as it traveled tremendous distance greatly increases the wavelength of the radiation, including the wavelength of the peak emission. An increased wavelength amounts to a decrease in the temperature of the radiation. Therefore, the big bang model predicts that the universe ought to be filled with blackbody radiation (the CMB) in the microwave part of the spectrum, with a temperature only a few degrees above absolute zero. This radiation field would appear to come from every direction. However, in

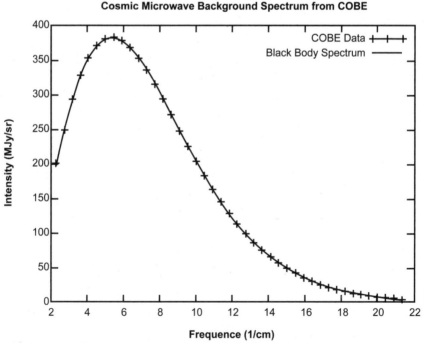

Figure 10.1. Blackbody spectrum of the CMB. (Wikimedia. Public Domain).

the steady state theory, the universe has always been as it now exists. Therefore, there never was a very hot, dense state in the universe, and so there cannot be a CMB. The 1965 discovery of the CMB (with a measured temperature of 2.73 K) contradicted the steady state theory and appeared to confirm the prediction of the big bang model. Consequently, most astronomers and cosmologists quickly abandoned the steady state model and embraced the big bang model. The big bang model has remained the one viable cosmology for a half century.

Changes in the Big Bang Model

It is questionable whether the status of the big bang as the sole cosmology has been a good thing for science. Over the years, the big bang model has faced many challenges. Each time, either the big bang model or ancillary theories have been altered in response to these problems. For instance, the estimated age of the universe (the time since the big bang) is inversely proportional to the Hubble constant. When the Hubble constant was 50 km/s/Mpc, we were confidently told that the universe was 16–18 billion years old. The range in the age was due to uncertainty in the role that the gravity of matter in the universe has had in slowing expansion. There was some tension here because the ages of globular clusters determined from their H-R diagrams were at least 15 billion years. However, when the Hubble constant was revised upward in the early 1990s, it required decreasing the age of the universe to well below the established ages of globular clusters. This was a problem because it meant that some things in the universe were older than the universe itself. Ultimately, astronomers reevaluated the ages of globular clusters, decided that they were in error, and revised them downward accordingly.

Cosmologists assume that the gas of the big bang was uniformly distributed. However, as a uniformly distributed gas expands, it cannot condense into the structures that we see in the universe today, such as galaxies and stars. Cosmologists propose that the gas in the early universe was not absolutely uniform, but that there were ever so slight variations in density. As the universe expanded, regions where the density was slightly greater would have acted as gravitational seeds to attract surrounding gas to form the structures of the universe. Incidentally, there must have been a delicate balance here, because if the gas were too uniform, no structure would appear in the universe, but if the early universe were too clumpy, most matter would have been gobbled up into black holes early on. Either way, we would not see the universe that we do today. Cosmologists determined how much clumping they needed in the early universe to account for the structure that we see today.

This slight clumpy distribution of matter in the early universe would have resulted in slight temperature variations that would show up in the CMB. Based upon their model of how much clumping was required, cosmologists predicted that if we precisely measured the temperature of the CMB in many directions, we would find temperature variations of one part in 10,000. Since the CMB has a temperature of about 3 K, the predicted temperature fluctuations ought to be 0.0003 K. Such precise temperature measurements of the CMB are difficult to perform from the earth's surface, so in 1989 NASA launched the COBE (*COsmic Background Explorer*) satellite to test the predictions. COBE was specifically designed to measure the predicted temperature fluctuations, and over its two-year mission, it measured the entire sky for the first time. It found that the CMB was perfectly smooth, contradicting the predictions. Scientists used a sophisticated statistical method to tweak out of the data slight temperature fluctuations an order of magnitude fainter. That is, instead of temperature variations of 0.0003, the researchers claimed temperature variations of 0.00003 K existed. This was beyond the ability of COBE to measure, so the research team could not produce a map showing where any of the temperature fluctuations were, but they assured us that they were real. At the time, this struck me as a very strange result, but additional studies with more sensitive equipment eventually confirmed the findings, and even produced maps of the temperature fluctuations. There have been two additional spacecraft launches with improved instruments to measure the CMB more precisely, the WMAP (*Wilkinson Microwave Anisotropy Probe*), launched in 2001, and *Planck*, launched in 2009.

Lost in all of this is the fact that the predictions were off by a factor of ten from the measured results. According to how the scientific method is supposed to work, this disproved the model, so the model ought to have been abandoned. However, the big bang was the only cosmological model, so it had to be retained. What was the fix? With the correct data in hand, cosmologists altered the big bang model to fit the data. From time to time, one hears the claim that the predictions perfectly matched the COBE measurements, but this is patently false. One can use the data to constrain the model, but it is improper to turn around and use the concordance of the two as some sort of proof of the model. All that concordance proves is that the modelers did a good job in altering the model to fit the data.

This sloppy confusion of the alleged confirmation of the model with the constraint of the model is all too common with the big bang. A similar thing happens with the abundances of the light elements. As mentioned in the previous chapter, the big bang universe supposedly began with the three lightest elements, with all other elements synthesized in stars. The conditions in the first three minutes after the big bang supposedly synthesized six isotopes,[2] two each of hydrogen, helium, and lithium. A common claim is that the big bang model correctly predicts the abundances of these isotopes, and since the measured abundances of those isotopes agree with the prediction, this is further proof of the big bang model. However, the abundances of the light elements were known prior to the prediction of the model. In reality, the measured abundances of these isotopes constrained the physical conditions within the first three minutes of the big bang. Hence, these isotopic abundances were not a prediction of the model but rather were constraints placed on the model.

There are additional problems for the big bang model that have arisen. As previously discussed, the temperature of the CMB is very uniform. Consider any two diametrically opposite points in the sky. They have the same temperature to within one part in 100,000. Why? It could be a coincidence that diametrically opposite points have the same temperature, but if we consider that the coincidence exists in every direction, the probability of this being the case is vanishingly small. If it is not coincidence that the temperatures are the same, then it must be that diametrically opposite points once were in thermal contact so that their temperatures equalized. Thermal contact means that two things can exchange heat through conduction, convection, or radiation. But how could thermal contact have been possible? Radiation from one point, say point A, is just now reaching our location. Radiation of the diametrically opposite direction, say point B, is just now reaching our location too. Since our location is halfway between points A and B, and radiation from either point is just now reaching our location, radiation could not have yet traveled between

---

[2] Defined in Chapter 3.

points A and B. This difficulty is called the *horizon problem*. The name comes from disparate parts of the universe being beyond the visible horizon from one another.

During the 1980s, cosmologists proposed cosmic inflation to solve the horizon problem. Cosmic inflation is a hypothetical rapid expansion in the early universe sometime between 10^{-36} and 10^{-32} seconds after the big bang. Cosmic inflation was far faster than the speed of light. Prior to cosmic inflation, the universe was exceedingly small, small enough to allow all the universe to be in thermal contact and hence come to the same temperature. However, cosmic inflation caused most of the universe to fall out of thermal contact, but a uniform temperature was preserved. A side benefit is that cosmic inflation solved another difficulty, the *flatness problem*. The flatness problem concerns the ratio of two energies, kinetic and potential, usually expressed by the Greek letter Ω. The value of Ω is very close to one, though, after billions of years of expansion, it ought to be either very close to zero or a very large number. Being so close to one looks very contrived, and hence improbable. However, cosmic inflation would have rapidly driven Ω almost identically to one, with only a modest retreat from one in the ensuing billions of years since. This is called the flatness problem, because with certain versions of the big bang model, Ω being one corresponds to flat geometry in the universe.

Because cosmic inflation solves the horizon and flatness problems so well, today few cosmologists or astronomers doubt its reality. However, there are at least two sticky issues. First, the mechanism of cosmic inflation is unknown. Why did inflation happen when it did? Why did it end when it did? And why has it not returned since? No one knows. Physicists usually describe phenomena like inflation with *fields*. A field is a property of space that permits interaction of some force. Examples of fields include gravitational fields, electric fields, and magnetic fields. In the case a gravitational field, it is the presence of matter that creates the field, and other matter within the field accelerate in response to the effect of the field. The presence of electrical charges creates electric fields, and the movement of charges creates magnetic fields. Those fields in turn accelerate other charges placed within the fields. Theoretical physicists have attempted to write a mathematical expression to describe a field that produces inflation, but there is no agreed upon form for this. The second issue is that there is no evidence that cosmic inflation has happened. Again, most astronomers and cosmologists now accept that inflation happened, not because of any evidence, but because without it, the big bang model is in serious trouble.

There have been many other alterations along the way. *String theory* is popular among particle physicists today. String theory proposes that there are many more dimensions than just the three of space and one of time.[3] Particles are vibrations in these extra dimensions. We can think of these as one-dimensional strings that vibrate, much as a string on a musical instrument vibrates. There are different versions of string theory, but theorists generally agree that the minimum number of extra dimensions required is six. We do not normally detect these extra dimensions because they are rolled up within matter very tightly. Theorists expect that at extremely high energies these extra dimensions can manifest themselves, but the energy required greatly exceeds those achievable by particle accelerators now in use. In a very early big bang universe, however, the energy (temperature) where strings would manifest themselves would have existed, so some versions of the big bang model now include string theory. Since there are no testable predictions of string theory, it is not clear that string theory is true. Including string theory in big bang models thus is piling conjecture upon conjecture.

As discussed in the previous chapter, there is good observational evidence for dark matter, though this rather radical proposal is controversial. If the universe primarily is governed by matter (gravity), then omitting most of the matter of the universe would be a serious deficiency. Therefore, most big bang models today include dark matter. Furthermore, like cosmic inflation, dark matter has proved handy to solve some cosmological problems. One problem is the formation of structure in the universe. The gravity of dark matter has been invoked to explain how galaxies and other hierarchal structures developed in the early universe. Another problem that dark matter supposedly solves is the perceived need for more matter than normal matter can provide.

In applying his new theory of general relativity to the universe a century ago, Albert Einstein found that his equations suggested that the universe was

---

[3] In general relativity, there is not a stark contrast between time and the three dimensions of space. Albert Einstein multiplied time, t, by the speed of light, c, to transform time into a sort of spatial dimension, ct, which he combined with the three spatial dimensions into a single equation. This is not so odd—we often express distances in terms of travel time, such as the distance between two cities being so many hours by a certain mode of transportation.

non-static. In the general case, being non-static means that the universe is either expanding or contracting. Apparently, Einstein only saw the possibility of contraction, not expansion. He thought that the universe was eternal, so contraction due to the gravity of all matter in the universe was unacceptable, because in an eternal universe there would have been ample time for all the matter in the universe to have accumulated into a single heap. To solve this dilemma, Einstein introduced the *cosmological constant*. The cosmological constant is a term that describes the self-repulsion of space. Einstein set the value of the cosmological constant to balance the effect of gravity to contract the universe, thus preserving an eternal universe. This is a static universe, with no contraction of expansion. Once others saw the possibility of expansion, and when Hubble confirmed expansion, Einstein retracted his cosmological constant.

Einstein was too harsh on himself, because the cosmological constant amounts to a constant of integration, a common thing in solutions of physics problems. The general practice is to set the constant of integration by the constraints of the problem. In many cases, the value of the constant of integration is zero, but one ought not to select that value simply because one does not know what the value of the constant ought to be. However, this is what cosmologists did for nearly 75 years. In 1998 and 1999, new data came forward that suggested that the expansion of the universe has accelerated. The most straightforward interpretation is that space is repelling itself, so the cosmological constant is not zero after all, though it does not have the value that Einstein preferred either (recall that he was dealing with a contracting universe, when in fact the universe appears to be expanding instead). However, modern cosmologists are hedging their bets. The acceleration in universal expansion could be a constant, but what if it is not? It is possible that the acceleration is not constant but instead may be a function of time or even space. Rather than preclude that possibility, the modern view is to call this acceleration in universal expansion *dark energy* rather than the cosmological constant. That choice was picked to make it somewhat parallel to dark matter. Like other forces, we can describe is as a field, and since a force can do work, it also can be viewed as energy (work and energy are equivalent), hence the other part of the name. The debate over the reality and nature of dark energy is presently in its infancy.

While the big bang model has been adapted to meet new challenges, there remains at least one problem that changes to the big bang model probably cannot fix. Because antimatter is a staple of science fiction stories involving space travel, many people think that antimatter is made up. However, antimatter is real. There are subtleties in the physics of particles that make matter and antimatter opposites of one another. For charged particles, the most obvious difference between the two is that the matter and antimatter equivalents have opposite charge. When a charged particle and its antiparticle come into proximity, their opposite charges attract one another. When the two charges meet, they annihilate one another, releasing their energy equivalent via the equation $E = mc^2$. Conversely, energy can convert into matter, but it will produce equal amounts of matter and antimatter. Therefore, the energy of the big bang ought to have produced symmetry between matter and antimatter. Very quickly, all the matter and antimatter would meet and destroy one another. Matter (not antimatter) dominates the universe, so clearly this has not happened. One could argue that portions of the universe are dominated by matter while antimatter dominates other portions. However, there would be boundaries between these regions where matter and antimatter would meet. We easily could see the annihilation of the matter and antimatter at those boundaries, but we do not see that. Physicists have no idea why the big bang universe produced asymmetry between matter and antimatter. This fundamental asymmetry in the universe probably indicates that naturalistic explanations of the origin of the universe are doomed to failure.

Let me recap by comparing the big bang model today with the big bang model of the early 1980s. The differences are as follows: (1) The Hubble constant then was much less than today, which means (2) the age of the universe posited then was less than is assumed today. (3) Big bang models then did not include cosmic inflation. (4) Big bang models then did not include dark matter. (5) Big bang models then did not include dark energy. And(6) big bang models then did not include string theory

In short, the big bang model of a few decades ago bears almost no resemblance to the big bang model today. For instance, into the 1990s we were confidently told that the big bang happened 16–18 billion years ago, but now we are assured that the big bang was 13.8 billion years ago, plus or minus 1%. Notice that the ranges of those two age estimates do not overlap, so at least one of them must be wrong. Yet supporters of the big bang model in the early 1980s had as much confidence as big bang supporters do today in the current model. What will the big bang model look like in a few more decades? We do not know, but we can be certain of two things: the big bang model then will be very different from today's version, and its supporters will have utmost confidence that it is correct.

In the estimation of many big bang believers, the big bang model has survived many challenges, and hence it is a robust theory. However, it owes its survival to the many fixes that it has been allowed. If an idea can be altered so easily in response to any new problems, how can it be disproven? A theory that explains anything and everything explains nothing. Having only one cosmological model for a half century is not a healthy thing, for the big bang model now has the trappings of dogma. This is not science.

Chapter 1 discussed how in the early second century Ptolemy explained the motion of the naked-eye planets with a system of epicycles. This model proved to be so successful in explaining planetary motion that it was the dominant cosmology for the next 15 centuries. In terms of longevity, the Ptolemaic model undoubtedly is the most successful theory in the history of science. Its strength was that the Ptolemaic model could be modified when new observations were at variance with its predictions. These modifications normally took the form of additional, smaller epicycles. However, this strength also turned out to be its weakness. The original Ptolemaic model had about a dozen epicycles to reproduce the motions of the five naked-eye planets, the sun, and the moon. However, by 1600, some versions of the Ptolemaic model required more than 100 epicycles. As also discussed in Chapter 1, application of Occam's razor eventually caused scientists to abandon the geocentric Ptolemaic model in favor of the heliocentric model. Many of the revisions to the big bang model resemble the addition of epicycles.

Is the Big Bang in the Bible?

Since the time of the ancient Greeks, it was commonly believed in the West that the universe was eternal. This probably was because the ancient Greek gods were not transcendent. Greek gods were born and could die. In many respects, those gods were not much more than supermen. As powerful as they were, the ancient Greek gods (as well as other pagan deities) were incapable of creation. Neither could the ancient Greeks conceive that the world created itself or otherwise arose out of natural processes. Therefore, it was much easier to believe that the world has always existed. Even as Christianity became popular in the West, with the Christian Scriptures proclaiming from the first verse that there was a beginning and that God created the world, this deep-seated notion of the eternality of the universe persisted. Thus, when the idea that the universe began with a big bang came along in the twentieth century, it was viewed as a radical concept. However, that part of the big bang model—that the universe was not eternal—ought not to have been so anathema to those who claimed to believe Scripture. Thus, once the big bang became widely accepted, some Christians and non-Christians[4] alike noted this one similarity between the Bible and the big bang. However, some Christians have gone beyond this, arguing that the big bang proves God's existence,[5] or that the big bang is taught in the Bible.[6]

How well founded are these claims? Let us start with the second one first, that the big bang is found in the Bible. There are several supposed indications of the big bang in the Bible. One is the sudden appearance of space and time in Genesis 1:1. It is reasoned that since the big bang was the sudden appearance of space and time, Genesis 1:1 tells us that the universe began with a big bang. While one might think that in retrospect after concluding that the big bang is how the world came to be, such is not evident from a proper exegetical treatment of Genesis 1:1. George LeMaitre, the Belgian theoretical physicist and Roman Catholic priest who around 1930 published his primeval atom theory (the precursor to the big bang model), thought that he saw in his model biblical creation. However, LeMaitre hardly had an orthodox view of Genesis, because he did not believe the first several chapters of Genesis to be historical narrative. To distill the creation account to one single essence, that the universe had a beginning, glosses over many details. One is hard pressed to find, among those who have seriously contributed to the big bang model, anyone who takes Genesis seriously, let alone anyone who professes salvation in the Lord Jesus Christ. If the big bang were so strongly taught in Scripture,

---

[4] A good example of an agnostic who wrote on this was Robert Jastrow in his 1978 book, *God and the Astronomers*, New York, New York: W.W. Norton & Company.

[5] One example is the philosopher William Lane Craig.

[6] One example is the astronomer Hugh Ross.

Bible-believing Christians ought to have been leading the way in developing that model. Is Genesis 1:1 a statement about the creation of the space of the universe? As I argue in Chapter 2 of the companion book, *The Created Cosmos: What the Bible Reveals about Astronomy*, Genesis 1:1 may be an example of introductory encapsulation, a summary, if you will, of the creation account, with details following in Genesis 1:2–31. Furthermore, in that chapter, I give reasons to believe that Genesis 1:6–8 (the Day Two account) is the record of God making the space of the universe. If this is correct, then that would preclude any possibility of the big bang, for the earth existed prior to the rest of the universe.

Another argument put forward that supposedly indicates that the Bible teaches the big bang is that at least eleven verses in the Old Testament speak of the stretching of the heavens. Presumably, this refers to the expansion of the universe. However, as I showed in Chapter 2 of the companion book, most, if not all, of these verses appear to be referring to a past event, and are not a reference to ongoing expansion today. The entity stretched out on Day Two appears to be a much better fit for the intended meaning of these verses. If that is the case, then these verses could not possibly refer to cosmic expansion today.

As attractive as seeing the big bang in the Bible may be, there are difficulties in the details. It is clear from Genesis 1:2 that the earth existed from the beginning, but that is not possible in the big bang model. According to the understanding of how the universe evolved after the big bang, the earth did not come into existence until more than nine billion years after the big bang. Furthermore, in a big bang universe, stars first formed within the first billion years, about eight billion years before the earth formed. But Genesis 1 makes it clear that the earth was created before the stars. While the evolutionary details of the development of things on earth may not be directly related to the big bang, belief in the big bang does morph into these other secular theories, which in turn collide with the Genesis creation account. For instance, in Genesis 1, plants existed before the sun, and birds existed before dinosaurs. To solve these problems and many more, one must be very clever with how one interprets the biblical text. This approach leads to an understanding of the text that one would never get from the Bible. This is eisegesis (reading into the text a foreign meaning) rather than exegesis (drawing from the text the author's intended meaning). Why is it that people want to reinterpret God's word in terms of what man claims rather than reinterpreting man's ideas in terms of what God says?

I understand the attraction that reading the big bang into the Bible has. If we read the Bible that way, and if science has established that the universe began in a big bang, then this amounts to scientific proof that the Bible is true. However, this is not the first time that those committed to Scripture read the cosmology of the day into the Bible. One glaring example of this happening before is the Roman Catholic Church's embrace of the Ptolemaic model during the Middle Ages, insisting that this was how God made the world, and so interpreted Scripture accordingly. This had disastrous consequences once the Ptolemaic model was rejected, because that led some people to conclude that the Bible was in error. Of course, it was not the Bible that was in error, but man's understanding of the Bible, based upon what science of the time dictated. We still live under the burden of that debacle in that many critics view the evolution/creation controversy through the lens of what supposedly happened four centuries ago at Galileo's trial.[7]

Another example of a popular cosmology being read into the biblical text is seen in how Hellenized Jews of the latter first millennium BC came to view the rāqîaʿ made on Day Two of the Creation Week. As I discussed in Chapter 1, as well as in Chapter 2 of the companion book *The Created Cosmos: What the Bible Reveals about Astronomy*, the Septuagint translators translated the Hebrew word rāqîaʿ into Greek as *stereoma*. Their motivation for doing this probably was to incorporate the dominant cosmology of their Hellenized culture. Immersed in a culture that held to the same cosmology, Jerome translated rāqîaʿ into Latin as *firmamentum*, which early English translations of the Bible, such as the King James Version, simply transliterated as *firmament*. Both Medieval Jewish and Christian theologians endorsed the ancient Greek cosmology, either as a lens to interpret Genesis or as an attempt to prove that the Bible is true. Of course,

---

[7] Far too many people draw the wrong lesson from the Galileo affair, thinking that it was an example of religion sticking its nose into a matter it had no business addressing. It was not the theologians who called for Galileo's censure, nor did Scripture play a major role in refutation to Galileo. It was Galileo's fellow scientists who cried foul for his challenging the scientific status quo, and most of the refutation came from Aristotle and Ptolemy. If one wishes to make a parallel to the evolution/creation debate, it is the creationists who are in the role of Galileo questioning the status quo, and evolutionists who are defending doctrine.

the ancient Greek cosmology upon which this was based was rejected long ago, but we live with the aftermath of this attempt to accommodate a two-millennia-old cosmology into Genesis. This has caused countless Christians to view the *rāqîaʿ* improperly, and hence has led to an improper view of biblical cosmology. Many people dismiss the reliability of the Genesis creation account because supposedly it reflects the false cosmology of the ANE with a domed, hard sky. It is ironic that both skeptics of the Bible and some Christians who want to uphold the integrity of Scripture make the same argument.[8] (Note that technically the cosmology the medieval sources endorsed was Greek, not that of the ANE. It is merely assumed that much earlier the Jews got their ideas about cosmology from the ancient cultures surrounding them.)

Finally, as will be discussed in Appendix A, there has been a recent resurgence of interest in the flat-earth cosmology among Christians. A major part of the claimed support for the flat-earth view are passages wherein the Bible purportedly teaches that the earth is flat with a dome above. However, this claim is based upon what people of the past claimed the Bible said, not what the Bible truly says. The people who have bought into this argument are recent extreme examples of victims of attempts to read cosmology of a different age into the Bible. In view of the lessons of the negative repercussions from previous attempts to wed the Bible to the prevailing cosmology of the day, ought we not be wary of repeating this mistake? If the history of science has any lessons, it is that scientific theories eventually are discarded in favor of something new. When that happens to the big bang model, what is to become of the integrity of Scripture?

Let us return to the other claim, that the big bang proves God's existence. When first proposed, the big bang model met with much resistance. The reason for the resistance was twofold. First, positing that the universe had a beginning ran counter to two millennia of belief in an eternal universe. Such deeply ingrained beliefs are not discarded easily. Second, to many people, the idea that the universe had a beginning suggested that there must have been an agent that brought the universe into existence. That is, there must be a God. Once the big

---

[8] An example of a Christian who makes this case may be Peter Enns. Enns basically argues that since the Genesis creation account reflects the incorrect cosmology of the ancient world, the details of the first few chapters of Genesis do not matter. What matters is that God made us and the world.

bang became firmly established, most scientists either set aside that objection or, as we shall see in the next section, attempted to circumvent it in various ways. Some Christians took up the cause and embellished the argument. An example of this is the philosopher William Lane Craig's development of the *Kalām* Cosmological Argument.[9] A syllogism central to Craig's argument is:

> Major premise: Whatever begins to exist has a cause
> Minor premise: The universe began to exist
> Conclusion: Therefore, the universe has a cause

Building on this premise, Craig went on argue for certain properties required of the cause, reasoning that those properties, such as transcendence and power, match those of the God of the Bible. As you might expect, other philosophers, as well as physicists, have disagreed with Craig. I will not repeat their objections here. Rather, I will briefly discuss what I see as deficiencies in Craig's argument.

What is the basis for Craig's major premise? It is based upon the observation of countless cause and effect relationships. In every case, we see that a cause precedes its effect in time. Though it may be hypothetically possible for a cause and its effect to be simultaneous, I cannot think of a single example. However, we can be certain that no effect can precede its cause in time. This temporal aspect of the causality argument is key in unpacking a few hidden assumptions in Craig's argument. Many people conceive of the big bang as an event in time. That is, they believe that time has always existed, and thus there was time before the big bang. If this were the case, then one *could* use the syllogism above to argue that the cause of the big bang preceded the big bang in time. However, since that cause was bound by time, the cause is not transcendent of time, and so falls short of the nature of the biblical God. Therefore, the assumption that time existed prior to the big bang fails to give a convincing argument that the God of the Bible exists.

Beyond that, however, belief that time preceded the big bang is a common misconception. The big bang model does not posit that matter and energy suddenly appeared in space and time. Rather, the big bang model requires the sudden appearance of matter, energy, space, *and* time. That is, space and time

---

[9] See William Lane Craig, *The Kalām Cosmological Argument* (London, United Kingdom: McMillan Press, 1979).

abruptly appeared with the big bang and have no antecedent. The big bang corresponds with $t = 0$, and there is no negative time. Sometimes people ask, "What was here before the big bang?" Within the big bang model, that question makes no sense, because *here* was not here then, and *then* was not then, then. The word *here* denotes space, and the word *then* denotes time, but neither existed until the big bang. All our experience with causality works within time, but for Craig's argument to work, causality must operate through the barrier of when time came into existence. One could argue that there is a causality argument that operates across that barrier, of which the causality that we see operate is more limited within time, but that would be asserting something that we do not know to be true.

Either Craig does not realize this flaw in his causality argument, or he does not grasp what the big bang model teaches about the origin of time. If the latter, then Craig effectively has substituted his own version of the big bang for the standard big bang model. Certainly, many Christians who agree with Craig's *Kalām* cosmological argument do not understand this point. Hence, this argument plays upon the ignorance of people, for it is a bit dishonest to claim that the standard big bang model has been proven, and then to covertly slip a *different version* of the big bang model into its place to prove God's existence. Nearly everyone agrees that medieval scholars' numerous attempts to prove God's existence with causality arguments failed. Is it not arrogant to think that today, finally, we are the first to succeed in making this case?

Where Did the Big Bang Come From?

Of course, one can be correct about the role of God in creating the universe, even if a rigorous proof of that proposition is not possible. But where does that leave the atheist regarding the question of where the big bang came from? There is no satisfactory answer to that question, though there has been much time spent upon it. By the early 1970s, some physicists were suggesting a possible answer from an interpretation of quantum mechanics. Quantum mechanics is one of the twin pillars of modern physics, and the other is general relativity. While general relativity describes matter and energy and how they interact with space and time on the largest scales, quantum mechanics describes the smallest systems, atoms and subatomic particles. Quantum mechanics describes particles with waves. When squared, the wave equation of a particle provides a probability function that does a remarkable job in describing the behavior of the particle. Where the probability is high for the location of the particle, we find that the particle most often is there. Where the probability is low, we do not find the particle there as often. The predictions of the probability function and the data agree well.

This prediction of the probability of locating a particle at a given location is very different from the deterministic prediction of classical mechanics, which gives us the precise location with absolute certainty. This means that in the world of quantum mechanics, there is a level of uncertainty, or fuzziness, that does not appear in the world of classical physics. This fuzziness comes about from the fact that, by their very nature, waves are hard to localize. When most of us learned in school that the electron orbits the nucleus of the atom, we pictured a tiny ball orbiting a larger ball. At least in our minds, the electron could be localized to a point on an orbit at any given time. However, this notion is naïve. Properly described in quantum mechanics, the electron is a wave that forms a cloud around the nucleus. Where the density of the cloud is greatest, the electron is more likely to be found, while the electron is less likely to be found where the cloud is less dense. While this picture may sound bizarre to many people, experimental results are consistent with it.

One way of expressing the inherit fuzziness of the quantum world is with the Heisenberg uncertainty principle. One form of the uncertainty principle is

$$\Delta E \, \Delta t \geq \hbar/2,$$

where ΔE is the uncertainty in energy, Δt is uncertainty in time, and $\hbar$ (called "h-bar") is Planck's constant (h, a value used in quantum mechanics) divided by 2π. What does this equation mean? If we enquire about the energy and time aspects of a quantum mechanical phenomenon, there is fundamental uncertainty in our knowledge of the two quantities. As we increase the precision of our knowledge of one quantity, thus lowering its uncertainty, we do so by increasing the uncertainty in the other quantity. Unlike classical physics, we cannot know both quantities with definite precision. This uncertainty is fundamental—it has nothing to do with experimental errors. Contrast this to classical physics, where definite precision is at least hypothetically possible, even though precision is practically limited by our ability to measure so precisely. As odd as this sounds, there are experimental results that are consistent with this understanding. The value of $\hbar$ is very small, so it does not show up in the everyday world, but its value is large enough to show up in quantum mechanical systems.

Some physicists have taken the uncertainty principle a step further. They argue that since the uncertainty of quantum mechanics is fundamental, the uncertainty principle can lead to violations of the conservation of energy. The violation of energy is equated with an uncertainty in energy. In this view, the violation of conservation of energy is permitted, provided the length of time that the violation exists is short enough so that the product of the energy violation and the time is approximately equal to $\hbar$. With this interpretation, particles briefly pop into existence and then cease to exist. These *virtual particles* have energy corresponding to their equivalent mass via the $E = mc^2$ equation. For the product of the energy and time involved to be approximately $\hbar$, the particles can exist for only a very brief time. While this understanding has become the majority opinion of physicists today, some physicists dissent.

These supposed possible violations of conservation of energy are called *quantum fluctuations*, and cosmologists have suggested that the universe began this way. *If* quantum fluctuations are real, and *if* the total energy of the universe is zero, then the universe could exist forever, or at least for a very long time. For this to work, both propositions, that quantum fluctuations are real, and the energy of the universe is zero, must be true. As already indicated, there is some question about the first proposition, but how possible is the second proposition? It does not appear to be possible. The universe contains a tremendous amount of energy in the form of matter $(E = mc^2)$ and photons of light (the energy of a photon is $E = h\upsilon$, where υ is the photon's frequency). Both forms of energy are positive, so for the energy of the universe to be zero, there must be a tremendous amount of counterbalancing negative energy. In physics, the only negative energies are potential energies. Potential energies are energies due to the location of an object. Physicists find it useful to describe potential energies with fields. An example of a potential energy is gravity; we describe it with a gravitational field. However, in classical physics, only differences in potential energy are significant. To handle this, a potential energy is measured with respect to a reference point where we assign the value of zero to the potential energy. The selection of the zero point is arbitrary—we pick it for convenience. Therefore, there is no true negative energy.

To counter this objection, it is claimed that the situation is different in general relativity, that the choice of the zero points of potential energies is not arbitrary, and hence truly negative energies can exist. Even if this is true, it does not guarantee that any fields present in the universe can provide sufficient negative energy to balance the tremendous positive energy that we all agree exists in the universe. Merely asserting this possibility falls far short of demonstration. Indeed, no one has even come close to showing that this is the case. Instead, all sorts of hypothetical fields in the early universe have been proposed to make this happen. Many of these dreamed up fields supposedly dumped tremendous energy in the form of matter and radiation into the very early universe as it expanded. A good discussion of this thinking is in Lawrence Krauss' 2012 book, *A Universe from Nothing: Why There Is Something Rather Than Nothing*. As the book's title suggests, Krauss thinks the universe sprang from nothing. Krauss' book is filled with conjecture, as indicated by the frequent use of tentative words such as *could*, *might*, and *may*. Krauss and others have taken very speculative ideas about physics, spun them into a tale about the origin and early history of the universe, and sold this tale to the public as a viable mythology upon which we can hang our eternal destinies.

None of this has been demonstrated physically, and it is questionable if it ever can be. Therefore, one must question whether any of this qualifies as science. Furthermore, for this hypothesis to work as claimed, it would require that something exist already, something where quantum mechanics applies so that a quantum fluctuation can occur. Krauss is aware of this, for he very cleverly discusses the nature of nothingness, arguing that the common conception of nothing actually is something. He then proceeds with a description of what true nothingness is. Are you confused? That is okay, because it may be Krauss' intent to confuse the matter. If the universe sprang from true nothingness, then quantum mechanics did not yet exist, and so there was no possibility of a quantum fluctuation. The best that one can say is that the universe came into existence from nothing in a manner that was consistent with itself. But again, this hardly is science.

As bizarre as this idea may sound, it gets worse. Let us go back to the wave equation that quantum mechanics gives as a description of particles. Physicists have always found the inherent fuzziness of the wave equation and the probability understanding of the wave equation unsettling. When quantum mechanics was developed in the mid-1920s, physicists almost immediately asked, "What does this mean?" The dominant school of thought since has been the *Copenhagen interpretation*, so called, because Niels Bohr and Werner Heisenberg developed it in Copenhagen. When we devise an experiment to measure the behavior of many particles subject to quantum mechanics, we find that they collectively behave as a wave. However, if we conduct an experiment where we detect the behavior of a single particle, it acts as a particle, not a wave. This is contradictory behavior in classical physics, because a particle cannot be both a wave and a particle. Furthermore, the experimental results suggested that if we do not detect individual particles, they collectively act as a wave, but when we measure just a particle, it clearly is a particle. How can this be? How could our observation of a particle change the outcome? The Copenhagen interpretation accepts the act of observation as significant. Before we observe the result of a single particle, it exists in all possible states dictated by its probability function. It is not until we observe the particle that it assumes one of the states permitted by the probability function. Physicists say that the wave collapses to assume that state.

While the Copenhagen interpretation of quantum mechanics quickly became the dominant opinion among physicists, it always has been controversial. For instance, Erwin Schroedinger, a key figure in the development of quantum mechanics, never accepted it. To illustrate his difficulty with the Copenhagen interpretation, Schroedinger developed the thought experiment that we call *Schroedinger's cat*. Suppose that we place a cat in a sealed box, along with vial of poisonous gas, and a hammer that can smash the vial, and thus kill the cat. The hammer is activated by a device, such as a Geiger counter, that registers the decay of an atomic nucleus, a process that is subject to quantum mechanics. After sufficient time, there is a certain probability that the cat is still alive, and a probability that the cat is dead. However, the cat is neither dead nor alive, but instead is in a metastate of being both alive and dead, until someone opens the box to examine the cat. At that point, the Copenhagen interpretation dictates that the wave function that describes the experiment collapses, and the cat assumes a state of being either dead or alive. As absurd as this sounds, we can take it further. Perhaps the scientist observing what state the cat is in is now part of the experiment, and thus he is a metastate himself, requiring that a second person observe him. Of course, that person now is part of the experiment, requiring, well... you clearly can see that this ends with an infinite regression.

What was Schroedinger's answer to the meaning of quantum mechanics? He suggested that the different probabilities were not alternatives, but that all possibilities played out simultaneously. In his *many worlds interpretation*, each possibility became a reality, defining an alternate universe. Since at every instant there are numerous quantum mechanical effects within our universe, over time our universe has spawned an unimaginable number of other universes. Again, this is hardly science. More properly, it is philosophy, though not particularly good philosophy. Before going on, the reader may ask why physicists ask these questions. This is an excellent question. For a long time, I have thought that the Copenhagen and many worlds interpretations of quantum mechanics are stupid answers to a stupid question. When Isaac Newton developed classical physics, no one asked what it meant. Everyone understood that it was a theory that adequately described the world as it exists. Quantum mechanics does the same thing, so why do we ask a question about what it means when we do not ask the same thing of classical mechanics? How quantum mechanics does this may seem strange to us, but that may be because of baggage that we carry from classical physics. When classical physics developed in the seventeenth century, it was difficult to comprehend for those hung up on Aristotelian physics.

In recent years, the many worlds interpretation of quantum mechanics has been invoked to explain the origin of the universe. If there are many other universes, with each new one birthed by a previously existing universe, then our universe is just one of probably an infinite number of universes. Each universe in this *multiverse* would be unique. Just as our universe began as a quantum fluctuation in another universe, quantum fluctuations in our universe often give rise to new universes. Each new universe immediately separates from its parent universe, becoming independent. How long has this cycle been going on? Again, the question of time outside of our universe does not make sense, but it is very clear that the multiverse is an infinite regression of universes both into the future and into the past. Therefore, the proposal of a multiverse amounts to a return to an eternal universe—albeit it is the multiverse that is eternal, while our universe itself is not eternal. On so many fronts, cosmologists who seriously suggest this are trying to have their cake and eat it too. In their view, time is pushed beyond the limits that it legitimately can be discussed, and a supposedly natural eternal entity is invoked to explain the origin of the universe. But this is hardly science, though the public may be swindled by this view, because scientists are the ones talking about it.

On this last point, there has been discussion as to whether the birth of the universe within a multiverse affected its structure and hence bears some mark of that process. If so, then some cosmologists argue that we might see an imprint of this in the CMB. Even if such a thing were possible, there could be many other mechanisms that could produce the same signal in the CMB. By definition, science is the study of the natural world. Also, by definition, other universes are beyond the natural world. This line of reasoning clearly is an attempt to bring something beyond the realm of science into the realm of science.

Perhaps what makes the multiverse attractive to so many people is that it dismisses a thorny problem for the big bang. There are many properties of the universe that make it ideally suited for man's existence. There are many examples of what we call the *anthropic principle*. For instance, if the fundamental constants that determine the structure of matter were slightly different, the chemistry of life would not be possible, and we would not be here to ponder this fact. When taken together, these coincidences appear to make the universe and our existence extremely improbable. Thus, the universe looks as if it were designed, which implies that there must be a Designer. People committed to naturalism strongly oppose this implication, and they loudly protest that such a conclusion is anathema to science. Their solution is to peddle nonsense such as the multiverse, all the while claiming that it is legitimate science. G. K. Chesterton beautifully summed up the futility of this kind of thinking:

> It is absurd for the Evolutionist to complain that it is unthinkable for an admittedly unthinkable God to make everything out of nothing, and then pretend that it is more thinkable that nothing should turn itself into everything.[10]

Romans 1:21 warns that when men forsake the knowledge of God, they become futile in their thinking and their foolish hearts are darkened. Man's desperate attempts to explain the world apart from God illustrate this better than anything else.

---

[10] G. K. Chesterton, *Saint Thomas Aquinas* (New York, New York: Doubleday Image, 1933), 174.

Biblical Cosmology

The cosmological ideas discussed in this chapter clearly were developed apart from any belief in God and without any regard for Scripture. Many Christians attempt to find biblical creation in these ideas, but should not the Christian begin with the Bible? In Chapter 2 of the companion book, *The Created Cosmos: What the Bible Reveals about Astronomy*, I discussed what I think is a proper biblical view of cosmology. I also discussed the topic in Chapter 1 of this book. Rather than repeat what I already written, I will summarize the key points.

The *rāqîaʿ* that God made on Day Two encompasses what we today would call outer space. On Day Four, God placed the astronomical bodies in the *rāqîaʿ*, and so one of the purposes of the *rāqîaʿ* is to be a place for the heavenly bodies. However, the primary purpose for the *rāqîaʿ* given in Genesis 1:6–7 was to separate the waters below from the waters above. Clearly, the waters below are the surface and ground water of the earth, but what of the waters above? If the *rāqîaʿ* is the space of the universe, then there must be water at the edge of the universe. All of this leads to three startling conclusions.

First, there is an edge to the universe. While permissible within the mathematics of cosmological models today, most cosmologists resist such a possibility.

Second, if the universe has an edge, then the universe is finite in size, and it has a center. The word *rāqîaʿ* refers to something that is spread or stretched out. If this spreading was reasonably symmetrical about the earth, then the earth is near the center of the universe. Again, while not contrary to the mathematics and physics of cosmological models, most cosmologists resist the idea that the universe has a center, let alone that the earth could be near that center.

Third, a layer of water is at the edge of the universe.

Again, water at the edge of the universe probably is not a problem per se, though most cosmologists would reject that possibility. If there is water at the edge of the universe, might not the cool, low pressure environment of space cause it rapidly to disperse or transform into solid or gas? If not, then what processes would prevent that? I do not know. For now, let us assume that water in some form exists at the edge of the universe. This layer of water forms a shell at least roughly centered on the earth. Being made of matter, the water must radiate. Depending upon the thickness and other physical conditions of this layer of water, that radiation would take the form of a blackbody spectrum. If most galaxy redshifts are cosmological, this water at the edge of the universe would have high redshift too, and so its blackbody spectrum would be redshifted as well, effectively cooling its temperature. Therefore, this model requires that there be a radiation field consisting of a cool blackbody coming from every direction of space. This is what we see in the CMB. The CMB is the one prediction of the big bang model, which is why the big bang became the single favored cosmology once the CMB was discovered a half century ago. For many years, biblical creationists have criticized the big bang model without offering any mechanism for the CMB. However, this biblically based cosmological model explains the CMB in a straightforward way, and so we now have a viable explanation for the CMB.

Conclusion

In the preceding chapters of this book, we have discussed at length the makeup of the cosmos, progressively journeying, as it were, from the earth, to the moon, to the planets, to the sun, to the stars, to the realm of intergalactic space. In the last chapter, we introduced the subjects of cosmology (the study of the structure of the universe) and cosmogony (the study of the origin and history of the universe). As we saw, scientists who assume a purely naturalistic origin for the universe have held tightly to the big bang model, continually updating, refining, and altering the model when it fails properly to account for all the observed data. Secular scientists have also proposed naturalistic models for the origin of the sun, the solar system, the earth, and the moon. Because we are dealing with historical science (which concerns scientific explanations for events that occurred in the past), the proposals associated with these models are not subject to direct testing. Even so, the observations we make, experiments we conduct, and mathematical equations we produce in the present reveal that many—if not all—of the proposed naturalistic models are implausible, if not downright untenable. The considerable number of variables which must be "just so" for naturalistic theories of cosmogony to work are staggering. Add to this the many speculative ideas about physics that must be relied upon to make these naturalistic theories of cosmogony even remotely viable, and they begin quickly to fall into the realm of mere wishful thinking. And indeed, this is what such naturalistic theories are— wishful thinking of individuals who, because they suppress the truth in their unrighteousness (Romans 1:18), are willing to believe any theory that excludes a Creator God. By embracing the current mode of scientific thought that denies the existence of God and allows for nothing beyond the material universe, they effectively condemn themselves to foolishness (Romans 1:21; cf. Psalm 14:1; 53:1). Arguably, it takes a great deal more faith to believe in some of the naturalistic cosmological theories than it does to believe in the existence of an all-powerful Creator who made the universe in the course of six days. However, for the devoted secularist, his assumption dictates his conclusion: The big bang model (or something like it) must be true because God *cannot* exist.

This assumption and conclusion leaves one in the position of being utterly arbitrary when it comes to answering the ultimate question about the existence of the universe. The ultimate question goes beyond considerations of the universe's temporal beginning, its formation, or its structure. Rather, it concerns purpose (i.e., teleology). Why does the universe exist? For the proponent of naturalism, there is no satisfactory answer. The universe exists because of a quantum fluctuation occurring in the midst of virtual nothingness that ushered in a 13.8-billion-year process of cosmic evolution. Along the way, random, unguided physical processes produced conditions that were just right for the formation of stars, galaxies, the solar system, and planet earth— including all of earth's life-sustaining systems. Life itself, perhaps not unique to planet earth, is not particularly special, but is the result of unguided chemical combinations and biological evolution. Eventually, the entire universe is destined for heat-death, the complete deterioration of all physical systems, and the attainment of maximum entropy. All life will have ceased to exist long before this time. The ultimate end of the universe is one completely devoid of purpose. As such, the whole naturalistic outlook on cosmology is one which fails to answer the question, "*Why?*"

The biblical view of cosmology, however, presents us with a definite purpose for the existence of the universe—a purpose that exhibits both theological and anthropological aspects. As we saw in the early chapters of this book, God created the sun and the moon for the unique purpose of giving light on the earth (Genesis 1:15, 17)—which was absolutely necessary for the existence of life. Similarly, the sun and moon were, along with the other celestial bodies, created for the purpose of marking the division between day and night, for marking the progression of time ("days and years"), for indicating appointed times ("seasons"), and for serving—on unique occasions—as signs related to the Lord's miraculous workings (Genesis 1:14, 18). Thus, the biblical text presents a clear picture of the purpose of the heavens in God's preparation of a world fit for mankind.

Beyond this, however, the heavens and all they contain were made for the express purpose of magnifying the glory of their Creator. Psalm 19:1 says,

> The heavens declare the glory of God,
>> and the sky above proclaims his handiwork.

The heavens, this verse says, "declare the glory of God," that is, in a nutshell, God's marvelous power, His unfathomable wisdom, and, ultimately, His deity (cf. Romans 1:20)—which entails a worthiness of honor and worship. The sheer vastness and grandeur of the heavens only serves to amplify this fact. Job 22:12 states,

> Is not God high in the heavens?
>> See the highest stars, how lofty they are!

While it seems unlikely that the author of Job could have fully grasped the incredible size of the heavenly expanse (indeed, we in our day have yet to fully get our minds around the size of the universe), he surely knew that the skies were of seemingly endless depth, with many of the celestial objects so distant that they were beyond any hope of man reaching them. The Lord, to have made outer space so vast, must have at His disposal unimaginably great power and unmatched wisdom. This fact is underscored by Scripture's reminder to its readers that God was unaided in making the cosmos. Everything we observe in the night sky, from the most intricate details of Saturn's rings, to the artistic grandeur of the Milky Way, was the result of God acting according to His perfect wisdom, mighty power, and sovereign design. Isaiah 44:24 records the Creator's words:

> Thus says the Lord, your Redeemer,
>> who formed you from the womb:
> "I am the Lord, who made all things,
>> who alone stretched out the heavens,
>> who spread out the earth by myself."

Human beings, including the most intelligent and innovative among us, are incapable of fathoming the essentially limitless power that must have been required to fashion the heavens. Even working together in large teams, we struggle to engineer things which pale by comparison to even a single star, much less a galaxy full of them. As such, it is particularly striking that Psalm

8:3 employs an anthropomorphism, describing the creation of all the stars as being merely the work of God's fingers. This ought to make us pause in humble, reverent awe of the Lord's omnipotence and omniscience that was brilliantly displayed in His work of creating the universe. But perhaps what is most striking about the Psalmist's description is that while the stars are described as merely being the work of God's fingers, the biblical text elsewhere emphatically states that God bares His mighty arm to accomplish salvation for His people (Isaiah 52:10; 53:1). God's creation of the stars and crafting of the galaxies, amazing though it was, pales by comparison to His work of redemption for those willing to receive His grace. The God who displays unfathomable might in creation exercises His power even more in bringing salvation to those who trust in Him.

This necessarily leads to a pivotal question. You see, even if you have concluded that naturalistic explanations for the origin of the universe are unsatisfactory, believing in a Creator is not enough. If God has created the universe and everything in it—including us—then He truly is a mighty God (Romans 1:20). As the sovereign Creator, God has a rightful claim on our lives, so it behooves us to inquire as to what that claim entails. Scripture tells us that God's requirement of us is absolute righteous perfection and holiness (Matthew 5:48; 1 Peter 1:15–16), just as He Himself is perfectly righteous and holy (cf. e.g., Psalm 71:9; 99:3, 5, 9; Isaiah 5:16). Mankind, however, comes pitifully short of God's perfect standard. Indeed, Romans 3:23 says that "all have sinned and fall short of the glory of God." Genesis 3 tells us how the disobedience of Adam (the first man) introduced sin into the world, and how through sin came death (cf. Romans 5:12). Romans 6:23 is unequivocal in its assertion that the "wages [or consequence] of sin is death." This death is both physical and spiritual. Ultimately, spiritual death results in eternal separation from God (cf. 2 Thessalonians 1:9). Thus, it is sin that prevents us from enjoying eternal life with God; however, God has provided a way to overcome sin.

God sent his only, unique Son, Jesus, into the world (John 1:1–18; 3:16), conceived by the Holy Spirit and born miraculously to a virgin named Mary. Jesus Christ is the perfect son of God. He lived a perfect, sinless life (2 Corinthians 5:21). He always did what pleased God His Father (John 8:29). However, Jesus voluntarily allowed Himself to be put to death by crucifixion. He died not for any sin He had committed; rather, Jesus died in the place of sinners. Isaiah prophesied Christ's death some seven centuries before the Crucifixion, saying,

> Surely he has borne our griefs
>> and carried our sorrows;
> yet we esteemed him stricken,
>> smitten by God, and afflicted.
> But he was pierced for our transgressions;
>> he was crushed for our iniquities;
> upon him was the chastisement that brought us peace,
>> and with his wounds we are healed.
> All we like sheep have gone astray;
>> we have turned—every one—to his own way;
> and the Lord has laid on him
>> the iniquity of us all. (Isaiah 53:4–6)

Notice that Christ's death was substitutionary: He was pierced for *our* transgressions; He was crushed for *our* iniquities. Similarly, 1 Peter 2:24 states, "He himself [Jesus Christ] bore our sins in his body on the tree, that we might die to sin and live to righteousness. By his wounds you have been healed."

As Jesus hung on the cross near death, He cried out, "It is finished" (John 19:30), a term used at that time to refer to a tax debt paid in full. Jesus Christ died to pay in full the debt to God that each of us owes. But because He is the sinless Son of God, death could not hold Him. Jesus rose from the dead just three days after His Crucifixion, thereby conquering death (cf. Acts 2:24). This is the *Gospel*, the good news of salvation for all people (1 Corinthians 15:1–4). Romans 4:25 teaches that Christ "was delivered over because of our transgressions, *and was raised because of our justification*" (emphasis added). Because of Jesus Christ's perfect life, sacrificial death, and conquering of death in His Resurrection, whoever believes on Him may be counted righteous in the sight of God (Romans 3:21–26). It is on account of Christ's atoning death that our sins may be forgiven (see Isaiah 53:10–12).

All that is required to receive salvation through Jesus Christ is to accept God's gracious gift by faith (repentant trust) in the Lord Jesus Christ (Acts 20:21). Romans 10:9 promises that "if you confess with your mouth that Jesus is Lord and believe in your heart that God raised him from the dead, you will be saved." Similarly, John 3:16 promises, "For God so loved the world, that he gave his only Son, that whoever believes in him should not perish but have eternal life." Jesus Christ Himself in John 11:25 says, "I am the resurrection and the life. Whoever believes in me, though he die, yet shall he live."

God's gracious gift of salvation is open to all. If you have not yet trusted in Christ for salvation, it is my prayer that you will do so today. For those reading this who do know the Lord Jesus Christ as their Savior, it is my hope that this book has encouraged you as you consider the power and wisdom of our God manifested in the heavens. I hope too that you might be inspired to praise our marvelous Creator for His awesome works.

> Praise the Lord!
> Praise the Lord from the heavens;
>> praise him in the heights!
> Praise him, all his angels;
>> praise him, all his hosts!
> Praise him, sun and moon,
>> praise him, all you shining stars!
> Praise him, you highest heavens,
>> and you waters above the heavens!
> Let them praise the name of the Lord!
>> For he commanded and they were created. (Psalm 148:1–5)

APPENDIX A

Is the Earth Flat?
Why Write About a Flat Earth?

Many people will probably wonder why it is necessary to defend a round earth, or, more specifically, an earth that is spherical. (You see, the earth could be both round and flat, if it were disk shaped.) Early in 2016, I had conversations with several people who were concerned about Christian young people they knew who were arguing that the earth is flat. One of these young people did not actually believe that the earth is flat. Rather, he found the topic interesting and the discussion of it stimulating. Indeed, it can be. In my years teaching at the university, I always asked the same sort of questions in my introductory astronomy classes to motivate my students into thinking more deeply. By raising the question, I challenged our cultural mythology that, until the time of Christopher Columbus five centuries ago, nearly everyone thought the earth was flat. Supposedly, with our sophistication and intelligence today, we know better than the ignorant people of the past. Most of my students were surprised to learn that the facts of history are very different. The question of the earth's true shape had been settled two millennia before Columbus. Also, rarely could any of my students give a good reason the earth is spherical. So much for our modern smug superiority over the supposedly ignorant people of the past.

Most people have not given this question any thought, because they have been taught their entire lives that the earth is spherical, so why worry about it? Consequently, with no idea of the reasons why we know that the earth is spherical, most people long ago entered a complacent state of more or less taking someone else's word for the matter. When someone comes along, such as this young man, who has given this some thought and begins to raise what appear to be simple objections to the earth's spherical shape, it doesn't take much to fluster most people. When cornered in this manner, people generally respond with the observation that we have photos from space that clearly show a spherical earth. My students usually came up with this answer too. However, I always pointed out that such photos easily can be faked. Indeed, because we all know that it is very easy to fake such photos, perhaps those photos don't prove much after all. Furthermore, those sorts of photos have been available only for a little more than a half-century. Belief in a spherical earth goes back much earlier than this, so obviously there must be better responses.

Once the arguments based on space photos of a spherical earth are shot down, the vast majority of people usually have one of two responses. The most common response is to dismiss the person asking the questions as a crank or fool, because "everyone knows that the earth is round." The other response is to pay more attention to the "flat-earthers," looking for errors in their facts or logic. However, rarely having the knowledge readily at hand to refute the case for a flat earth, most people who take this approach soon look for help. That search for help usually is on the internet, whereupon they quickly find a slew of websites and videos promoting the flat earth, but precious few, if any, refuting it. Some people emerge a few hours later, their egos bruised and their intelligence a bit insulted because they still think that the flat earth is nonsense but are frustrated that they can't seem to answer many of the arguments they've just encountered. Still others never emerge from this rabbit hole and end up thinking that maybe the conspiracy theories they have encountered along the way may be right. Perhaps for a long time we've all been fed a whopping lie about the true shape of the earth!

Despite the widespread belief in a spherical earth, at least in the West, for more than two millennia, there have been a few people who persisted in insisting that the earth is flat. The modern movement for a flat earth, though, has its origins in the nineteenth century. For much of the time since then, flat-earthers[1] have been viewed as cranks. However, there has been a resurgence of belief in a flat earth in recent years, especially among people who claim to be Christians. This appendix is intended as a response to this threat.

[1] The term "flat-earther" long has been used as a term of derision, so I was loathe to use it. However, I have noticed that some modern flat-earthers use the term to describe themselves in what appears to be an attempt to reclaim it. It is a more compact phrase than something like "those who support the flat-earth theory" to describe people who believe the earth to be flat. Therefore, I will use the term occasionally. Please understand that in this context I use it in a descriptive manner, not meaning it in an insulting way.

Reasons We Know the Earth Is Spherical

Chapter 1 discussed some of the reasons for believing that the earth is spherical, arguments that go back to ancient times. I will briefly describe some of those here as well. The earliest recorded discussion of a spherical earth is from Pythagoras. In the sixth century BC Pythagoras correctly understood that the cause of lunar eclipses is the shadow of the earth falling on the moon. This can happen only when the moon is opposite the sun in our sky, which coincides with full moon. The earth's shadow is larger than the moon, so we cannot see the entire shadow at once. However, during a lunar eclipse we see the earth's shadow creep across the moon. Because the edge of the earth's shadow always is a portion of a circle, the earth's shadow must be a circle. If the earth were flat and round, similar to a disk, it could cast a circular shadow, but only for lunar eclipses that occur at midnight. For a lunar eclipse at sunrise or sunset, the earth's shadow would be an ellipse, a line, or a rectangle, depending upon how thick the disk was compared to its diameter. However, the earth's shadow during a lunar eclipse is always a circle, regardless of the time of night when the eclipse occurs. The only shape that consistently has a circular shadow, regardless of its orientation, is a sphere.

Another argument involves the stars that are visible in the northern and southern parts of the sky. The North Star lies within a degree of the north celestial pole, the direction in space toward which the earth's rotation points. As the earth rotates each day, the stars, the sun, and the moon appear to spin around the north celestial pole, so the north celestial pole remains fixed in the sky. In the ancient world, many people thought that the celestial sphere rotated each day around a non-spinning earth. For our purposes here, it doesn't matter which is the case. The north celestial pole makes an angle with the northern horizon. We call this angle the altitude of the north celestial pole. Since the North Star is so close to the north celestial pole, we can approximate the altitude of the north celestial pole with the North Star's altitude.

The altitude of the North Star is noticeably higher in the sky at northern locations than it is at southern locations. For example, the North Star is much higher in the sky in the northern United States and Canada than it is in Florida, as anyone vacationing in Florida can attest (provided they pay attention). This can happen only if north-south motion is along an arc. This is further underscored by other considerations. There is a region around the North Star in which the stars do not rise or set but instead are continually up and appear to go in circles around the north celestial pole. We call these circumpolar stars, meaning "around the pole." The region of circumpolar stars is larger at northern locations than in southern locations. Likewise, there is a circumpolar region below the southern horizon whose stars are always below the horizon. The northern circumpolar region, where stars are always visible, is very large, and the southern circumpolar region, whose stars are never visible, is also large. Closer to the earth's equator, the two circumpolar regions are smaller. For example, for many years I lived in South Carolina, about 4° farther south in latitude from where I now live in northern Kentucky. I can see that the North Star is slightly higher in northern Kentucky than it was in South Carolina. Furthermore, during winter in South Carolina, the bright star Canopus barely rose above the southern horizon each night; but in Kentucky I can never see Canopus. This is because in northern Kentucky, Canopus is in the southern circumpolar region where stars are never visible, while in South Carolina it is not. This too shows that the earth is curved in the north-south direction.

Not only is the earth curved in the north-south direction, it also is curved in the east-west direction. There is a time difference of three hours between the east and west coasts of the United States. That is, the sun rises and sets approximately three hours earlier on the east coast than it does on the west coast. This is easily verified by anyone who has flown between the east and west coasts of the United States. Not only will your watch show that there is a time difference of three hours, but your body will notice the difference in time as well. If one drives from one coast to the other, the trip will take several days, so our bodies will not notice the time difference as much. However, our watches reveal that the time has changed. Such rapid transportation was not possible in ancient times, but the ancients could see this time difference another way. A lunar eclipse obviously must happen simultaneously for everyone on earth, but it will be different times at different locations. For instance, a lunar eclipse may start shortly after sunset in the eastern Mediterranean, such as in Greece. However, in the western Mediterranean, such as in Spain, the moon would already be in eclipse when the moon rose that night. This means that the lunar eclipse began before sunset/moonrise in Spain, but after sunset/moonrise in Greece. Communication in the ancient world was such that people were aware of this effect. This shows that the earth is curved in the east-west direction. If the earth is curved in both the north-south and east-west direction, the most likely shape of the earth is a sphere.

Ancient sources, such as Aristotle, also mentioned that the hulls of ships disappeared before their masts did as the ships sailed away. This would happen only if the earth is spherical. Without optical aid, this is difficult to see. However, one easily can see a related effect. If one is perched atop the mast of the ship, one can spot land or other ships before people on the deck can. This is why spotters often were placed in a crow's nest high above a ship's deck. If the earth were flat, there would be no advantage to being above the deck. A similar thing can be observed on land. The Door Peninsula in Wisconsin forms the eastern shore of Green Bay. The distance across Green Bay from the northern portion of the Door Peninsula to Northern Michigan is about 20 mi (30 km). Looking across Green Bay from the beach on the west side of the Door Peninsula, one cannot see Northern Michigan. However, if one ascends the bluffs above the beach, one can see the shoreline of Northern Michigan. This is possible only if the earth is spherical.[2]

Not only did ancient people know that the earth was spherical, one of them accurately measured the size of the earth around 200 BC. Eratosthenes worked at the Great Library in Alexandria, Egypt. Eratosthenes is the father of geography because he coined the term and commissioned the creation of many maps. One year on the summer solstice, Eratosthenes was in southern Egypt near modern-day Aswan. Being on the northern limit of the tropics, the sun was directly overhead at noon on the summer solstice. Eratosthenes realized this because he could look down into a deep well and see the bottom.

Normally, the bottom of a well is not visible because the sun's light does not shine directly on the bottom, but it did at noon on the summer solstice because the sun was directly overhead. The sun never was directly overhead in Alexandria, because Alexandria is not in the tropics. Back in Alexandria the following year, Eratosthenes measured the altitude of the sun at noon on the summer solstice. He did this by measuring the length of a vertical pole of known height at noon. Trigonometry allowed Eratosthenes to compute the sun's altitude. The difference between 90° and the altitude was how far the

---

[2] Even with a telescope, seeing the disappearance of the hull of a ship first often is difficult. This is because of a temperature inversion that frequently happens with air near the surface of a large body of water. This temperature inversion can bend the light from a distant object around the surface of the earth for many miles. However, if one ascends above this temperature inversion, as to the mast of a ship or to bluffs above a shoreline, and compares to the view at water's edge, the effect of the earth's sphericity is more apparent.

sun was from being vertical. Eratosthenes found that the angle was about $^1/_{50}$ of a circle. This meant that Alexandria and Aswan were separated by $^1/_{50}$ of the earth's circumference. Eratosthenes knew the distance between those two locations, so multiplying that distance by 50 gave him the earth's circumference.

Why do so many people today assume everyone thought the earth was flat until the time of Columbus? The argument at the time of Columbus was not over the earth's shape, but over the earth's size. Muslims had closed off to Europeans the overland trade routes to the Far East. Everyone realized that travel to Asia by sailing west from Europe was possible, but why would you want to? There was a vast ocean (they didn't know about the two American continents in between) separating Europe and Asia. It was much shorter to sail eastward from Europe, perhaps around Africa, to reach Asia. In the small ships used at the time, it was not advisable to sail more than a few days out of sight of land. Columbus was proposing a voyage of a few months over open, uncharted waters. That was very dangerous. To make his proposed voyage more palatable, Columbus overestimated the eastward distance from Europe, and at the same time he decreased Eratosthenes' measurement of the earth's circumference. The difference in these two was Columbus' expected distance to Asia by sailing westward from Europe. In Columbus' estimation, it was shorter to reach Asia by sailing westward than eastward. A glance at a modern globe or map of the world reveals that this is false. In other words, Columbus was wrong, and his critics were right!

In the late nineteenth century, two atheistic skeptics, Andrew Dickson White and John Draper, created the conflict thesis—that Christianity held back the progress of science. One of their major arguments was that, throughout the Middle Ages, the church had taught that the earth was flat. In creating this myth, Draper and White suggested that the church could redeem itself for this supposed error on the earth's shape by getting in on the ground floor of Darwinism. This ploy was very successful in that much of the church capitulated on evolution. It also falsely altered history. It is this false version of history that most people have learned.

The Recent Version of the Flat Earth

The most popular flat-earth cosmology promoted today is what some call a snow globe earth. A snow globe is a water-filled glass sphere on a base with a scene on a flat plate on the inside of the sphere on the sphere's bottom. Often the scene is one of winter. There are small white particles in the water that are slightly denser than water. If one shakes the globe and sets the globe down on its base, snow appears to fall down upon the scene for a while. The snow globe earth is flat and round, with a hard sphere above it. Instead of water, the sphere is filled with air. And there are no fake snow particles. The stars are embedded on the sphere above, and the sun and moon are either on the sphere or just inside the sphere. The center of the flat earth is the North Pole, with the continents and oceans distributed similarly to how they are projected on the United Nations flag.[3] Each day the sun (and presumably the entire sphere) spins around the North Pole, causing the sun and stars to move across the sky. There is no South Pole. Nor does Antarctica exist as we know it. Instead, the edge of the round, flat earth is ringed by a high ice wall that no one has been able to penetrate, so no one knows how far outward the ice extends. According to the snow globe earth model, since no one has penetrated far into Antarctica, no one knows where the hard dome of the sky touches the ground. Apparently, Roald Amundsen lied about leading the first expedition to reach the South Pole in 1911, as have all others who have claimed to have reached the South Pole since. Furthermore, the United States government lies when it claims that the Amundsen-Scott South Pole Station has been staffed continuously since 1956; and the many people who claim to have spent time working at the Amundsen-Scott South Pole Station are lying too. Sir Edmund Hillary and Sir Vivian Fuchs led the Commonwealth Trans-Antarctic Expedition, the first Antarctic land crossing (from one side to the other), in 1958 as part of the International Geophysical Year. Of course, if the flat-earth conspiracy theorists are correct, not only Hillary, Fuchs, and their crew, but every person who has claimed to have crossed the Antarctic continent since are liars too. The number of people supposedly involved in this vast conspiracy of the spherical earth is staggering.

The snow globe earth model is geocentric. That is, the earth remains stationary and does not orbit the sun each year. For that matter, the moon does

---

[3] In fact, some promoters of the flat earth suggest that the United Nations flag is an admission that the earth actually is flat. After all, it would seem that the people running the United Nations certainly would be in a position to know the earth's true shape. But this makes no sense—why would a conspiracy devoted to maintaining belief in a spherical earth create a flag that supposedly spills the beans on what is really going on? This reminds one of the supervillains on the campy 1960s television farce *Batman* who prominently displayed outrageous clues to their crimes.

not orbit around the earth each month, nor do man-made satellites orbit the earth. Man has not ventured into space, and the Apollo astronauts did not land on the moon. As you can see, belief in a flat earth naturally leads to numerous other conspiracy theories, such as the notion that the Apollo moon landings were faked. To be fair, there are people who deny that we actually landed on the moon who believe that the earth orbits the sun and that the earth is spherical. But many other people have come to believe that the manned space program and the moon landings are a hoax primarily because of belief that the earth is flat.

As bizarre as this modern flat-earth model may seem, it recently has gained considerable traction. In addition to the conversation that I had in early 2016 with several older people concerned about young people being taken in by the flat-earth movement, there were other indications of the spread of this new phenomenon. Several people who regularly speak about creation (including those who are full-time speakers for Answers in Genesis) reported that questions about the flat earth began occurring in 2015 and 2016. Answers in Genesis likewise began to receive correspondences enquiring about the flat earth. This immediately raised two questions: Who are the people responsible for this recent interest in a flat earth? And what is their motivation? It appears that many of the people interested in this question are young, suggesting that social media is a major conduit. A quick internet search reveals tens of thousands of hits on the web promoting a flat earth. There are some articles, but many of these are videos of varying length and quality of production. Some are very short, less than a minute long, but typically these videos last five to 15 minutes. Many videos are of rather poor quality, with inferior sound and graphics. However, a number of others are best described as documentaries that are well-done from a technical standpoint. Many of these documentaries run up to two hours long.

An internet search for material promoting a flat earth often seems like a lengthy trip through the rabbit hole. As in the story of Alice in Wonderland, not everything is as it appears. Most people are aware that there is a Flat Earth Society, thinking that the Flat Earth Society is a serious group of people dedicated to promoting their own peculiar view of the world. The situation is far murkier than that. Actually, there have been several Flat Earth Societies. Some of them clearly have been tongue-in-cheek, while others appear to be far more serious. Some flat-earth advocates obviously are having fun, and they don't seem to mind if their audience is in on the gag. However, some people promoting a flat earth appear to enjoy watching people squirm uncomfortably when confronted with an argument that they disagree with but can't quite manage to refute. Of course, these people are not about to let on that they are anything but serious about the flat earth. This is perverse.

Examples of Flat-Earth Proponents

An example of someone who may not be serious about the earth being flat is Matthew Boylan. According to some sources on the flat earth, Boylan is an artist who was an independent contractor with NASA. Furthermore, he supposedly left that job after NASA employees took him into their confidence and invited him to join the conspiracy promoting the lie that the earth is spherical. According to Boylan, NASA fakes nearly everything that it does. Boylan has several videos on the Internet, but some of them appear to be comedy routines. In at least a few of the videos, audience members certainly reacted in a manner that suggested they understood they were watching a comedy routine. Boylan's delivery, including his frequent use of profanity, is similar to many other comedy routines today. In this routine, Boylan included a photo that he says shows the Apollo 11 Lunar Excursion Module (LEM) landing on the moon. According to him, because there had to be a camera crew already on the moon to take the photo, NASA faked the moon landing. However, the photograph clearly shows the curved edge of the moon, indicating that the LEM was far above the lunar surface. Actually, this photo was taken by Michael Collins, who remained aboard the Apollo 11 Command Module (CM), as the two other astronauts, Neil Armstrong and Buzz Aldrin, landed on the moon. Collins took the photo shortly after separation of the LEM from the CM. Similarly, Boylan shows an image of the Galileo probe arriving at Jupiter. Boylan mockingly notes that the Galileo probe must have been followed very closely by another spacecraft carrying a camera. NASA frequently produces this sort of image of spacecraft superimposed over an image of some space object relevant to the spacecraft's mission. In other words, this is an artist's rendition. As an artist, Boylan must understand this. He must be very amused that so many people think that he is serious.

Probably the person who many flat-earth promoters consider most authoritative on the subject is Eric Dubay. Dubay has written at least two books on the flat earth, *The Flat-Earth Conspiracy* and *200 Proofs Earth Is Not a Spinning Ball*, and he is featured in, or is credited with, several videos on YouTube. From a production standpoint, these generally are among the better quality videos promoting a flat earth on the internet.

Dubay's argument (and hence the argument of those who have been influenced by him) for a flat earth relies upon a number of misunderstandings and false information. For instance, flat-earth arguments frequently repeat the previously mentioned cultural mythology that nearly everyone believed that the earth was flat until five centuries ago with the historic voyage of Christopher Columbus.[4] As previously mentioned, this faux history arose from the conflict thesis promoted by Andrew Dickson White and John William Draper during the latter part of the nineteenth century. Supposedly, the modern era was preceded by a long dark age after the fall of Rome, a dark age from which humanity had just recently emerged during the Enlightenment. The darkness of medieval times was caused by superstition perpetuated by Christianity. Once this impediment was removed during the Enlightenment, true progress could occur. One of the impediments supposedly was belief in a flat earth, which both the Bible and the church allegedly taught. It might make for a good story, but none of it is true. First, the Middle Ages were not quite as dark as often thought. Second, the church never taught that the earth was flat.[5] Furthermore, belief in a spherical earth has been common for more than two millennia. Both Ptolemy (early second century AD) and Aristotle (fourth century BC) taught that the earth was spherical. Not only did the teachings of Aristotle and Ptolemy dominate medieval thinking, but their teachings heavily influenced the medieval church as well. It would make no sense to accept all that these two men wrote, except for what they said about the earth's shape. There is no record that the church did; nor is there any record that the church taught that the earth was flat. There was not much discussion of the earth's shape in medieval writings because there was no question that the earth was anything but spherical. Supporters of a flat earth commonly claim that belief in a spherical earth accompanied a belief in evolution and a rejection of God

[4] Eric Dubay, *The Flat-Earth Conspiracy* (self-published, 2014), 3, 7, 242. Page numbers in future references to Dubay's work refer to this book.

[5] For a good refutation of the idea that the medieval church taught that the earth was flat, see the work of the medieval scholar Jeffrey Burton Russell, *Inventing the Flat Earth: Columbus and Modern Historians* (Westport, Connecticut: Praeger, 1991).

and the Bible beginning five centuries ago.[6] However, none of this stands up to historical scrutiny. A true student of history would know better than this.

On page 149 of *The Flat-Earth Conspiracy*, Dubay incorrectly stated that "the first person to ever present the idea of a Sun-centered universe was Pythagoras of Samos in around 500 B.C." Actually, it was Aristarchus of Samos, who lived about two-and-a-half centuries after Pythagoras, who proposed the first sun-centered system that we are aware of. Perhaps Dubay was thinking of Philolaus, a Pythagorean philosopher about a century after Pythagoras, who apparently was the first to propose that the earth was not the center of the universe. However, Philolaus' cosmology was not heliocentric, because he envisioned the earth, sun, moon, and other planets orbiting a central fire. Or perhaps Dubay has confused Pythagoras' conclusion that the earth is spherical with the heliocentric model. As previously mentioned, Pythagoras was the first person that we know of who taught that the earth is spherical.

Dubay's arguments also suffer from the all-too-common misunderstanding of the cause of the seasons. All of us were taught early in our education that the earth's 23½° tilt causes the seasons. As the earth orbits the sun each year, we alternately tilt toward the sun (resulting in summer) and away from the sun (resulting in winter). Unfortunately, most people fail to understand what this means, because most people, like Dubay, think that it is the changing distance from the sun due to the earth's tilt that is responsible for seasonal changes (pages 71–72, 245). It is not, as Dubay argues:

> if the heat of the Sun travels 93,000,000 miles to reach us, a small axial tilt and wobble, the difference of a few thousand miles, should be completely negligible.

If it is not the change in distance that causes the seasons, then what does cause the seasons? There are two effects at play. First, when tilted away from the sun, the sun's altitude is much lower in the sky than when we are tilted toward the sun. For instance, where I live the sun's altitude at noon on the summer solstice is a little more than 74°. At noon on the winter solstice, the sun's altitude is a little more than 27°. Because the sun's rays strike the ground at a lower angle in winter than in summer, the energy of the sun's rays is spread

---

[6] Eric Dubay in *The Flat-Earth Conspiracy*, 8, 172–215, discusses evolution and its supposed connection to spherical earth.

out over a much larger area than it is during the summer. In the specific case of my location, the ratio is a little more than 2:1. Since the sun's rays are required to heat more than twice the surface area in the winter than they do in the summer, winter is cooler. However, not only is the sun much lower in the sky during the winter, the sun is not in the sky very long during the winter. Again, at my location, the sun is above the horizon for nearly 14 hours and 45 minutes near the summer solstice, but the sun is up for a little more than 9 hours and 20 minutes near the winter solstice. With less sun exposure during the winter (and correspondingly more time for radiative cooling at night), winter is much cooler than summer. These two effects involving area and time combine easily to explain the seasons. Thus, in using this sort of argument against the spherical earth, Dubay exposes his ignorance of the true cause of the seasons. Whether intentional or not, this amounts to a straw-man argument.

But Dubay's ignorance is displayed in many other ways. He correctly states that if the earth revolves around the sun each year, the positions of the stars ought to shift back and forth. We call this effect parallax. Ancient astronomers debated this point, and because they failed to measure parallax, most concluded that the heliocentric theory must not be true. Some of the ancients who believed in the heliocentric theory responded that parallax would be too small to measure if the stars were incredibly far away. Indeed, that turned out to be the case. Parallax measurements require the use of a telescope. Even then, the first parallax measurements were not successful until the 1830s, more than two centuries after the invention of the telescope. Traditional parallax measurements from the ground are difficult, usually because of the blurring effect of the earth's atmosphere. Several spacecraft have made more precise parallax measurements possible. The Hubble Space Telescope can provide very precise parallax measurements, but only on a very limited basis. The HIPPARCOS mission in the early 1990s was optimized for measuring the parallaxes of many stars. The HIPPARCOS high precision catalog contains parallax measurements for nearly 120,000 stars. This produces distances with accuracy of about 20% for many of the stars within 600 light years of earth. The Gaia spacecraft, launched on a five-year mission late in 2013, will accurately measure the parallaxes, and hence distances, of millions of stars, increasing the margin for accurate distance measurements to about 6000 light years. Yet Dubay appears to be ignorant of this progress, for he wrote on page 12,

After almost two hundred million miles of supposed orbit around the Sun, not a single inch [*sic*] of parallax can be detected in the stars!

This claim is repeated on page 14 and on pages 242–243, and it is implied elsewhere. Many other flat-earthers have repeated this false information about the lack of parallax.

Dubay's book contains many false assertions, such as that the midnight sun cannot be explained by a spherical earth (page 70) or that Polaris, the North Star, can be seen as far as 23.5° south latitude (pages 71, 75) (Polaris generally is not visible south of the earth's equator). The only evidence for these false assertions—when any evidence is actually given—are quotes from various other authors writing in defense of the flat earth. And these quotes are frequent, often occupying entire pages of Dubay's book. Some of these other works are Samuel Rowbotham's 1864 *Zetetic Astronomy: Earth Not a Globe!*, William Carpenter's 1885 *100 Proofs the Earth Is Not a Globe*, Thomas Winship's 1897 *Zetetic Cosmogeny*, and David Wardlaw Scott's 1901 *Terra Firma: The Earth Not a Planet Proved from Scripture, Reason, and Fact*. Rowbotham's work is the original source on the flat-earth theory, and the other authors, as well as Dubay, have repeated and embellished much of what Rowbotham claimed. None of these are credible sources, so appeal to them hardly constitutes evidence.

In his videos and in his books, Dubay clearly demonstrates that he does not understand physics. In at least one video he claimed that rockets cannot work in space because there is no air. In actuality, rockets work because of Newton's third law of motion (action-reaction), not because they push off of the air. Dubay protests that gravity seems to have two contradicting properties: making things stick to the earth and causing other things to orbit the earth. If Dubay understood even elementary physics, he would know that because of Newton's first law of motion, an object requires a centripetal force in order to orbit. Gravity provides that force. This is no different from any other object that goes in a circular or nearly circular path. A weight whirled around by a string is compelled in its orbit by tension in the string. In a similar manner, gravity provides the force required to make the moon orbit the earth. Centripetal force required for circular motion is described fully in any physics course. From his discussion, Dubay clearly does not understand the Coriolis Effect. Different latitudes on earth have different radii from the earth's rotation axis, hence there are different rotation speeds at different latitudes. Consequently, as air currents move from one latitude to another, they deflect rightward on the earth's surface. This is the Coriolis Effect. The Coriolis Effect explains why the dominant wind directions are different at different latitudes, and why low-pressure systems spin counter-clockwise in the Northern Hemisphere (high-pressure systems spin clockwise, and the direction of either is reversed in the Southern Hemisphere). Incidentally, flat-earthers never even mention this, for there is no explanation in their model. There are many other examples of Dubay's failure to comprehend even basic physics. Many of Dubay's followers repeat his failure to comprehend physics.

Christians Supporting the Flat-Earth Belief

It is not clear who is primarily responsible for the spread of belief in a flat earth among Christians. There are numerous videos, both long and short, available on the Internet that appear to promote a biblical argument for a flat earth. One such resource is a relatively well-done documentary, *The Biblical Flat Earth Series: The Global Lie Flat Earth Revelation Documentary*. Three of the people credited in the documentary are Philip Stallings, Rob Skiba, and Robbie Davidson. Davidson apparently is the filmmaker and primarily responsible for the production of the documentary. Both Skiba and Stallings have other videos promoting a flat earth, as well as other ideas, on the Internet. Stallings is identified as the founder of the Bible Flat Earth Society. An organization called Celebrate Truth also was involved in the documentary. It is not clear what Celebrate Truth is or who is behind it. Both Celebrate Truth and the Bible Flat Earth Society appear to have a presence solely on social media.

This documentary repeats many of the false claims of Dubay, such as the claims that we do not see stellar parallax, and that stars are not nearly as far away as astronomers maintain. In addition, there are other details included that Dubay does not mention in his book. These include the denial of the existence of extrasolar planets, the denial that meteors strike the earth, and the denial that the source of the sun's energy is nuclear. This video also echoes the Aristotelian objection that if the earth were moving, it would leave its atmosphere behind. The video likewise posits a variation of this theme with the claim that an airplane could not land on a runway if the earth were moving. Supposedly, when a plane leaves a moving earth, the plane is left behind by the earth's motion. This false claim is made by other supporters of a flat earth.

Once one postulates a flat earth, it leads to other preposterous claims. If the earth is flat rather than a sphere, then it is inconceivable that we have ventured into space. In the previously mentioned interview with Eric Dubay, he denied that there are any satellites orbiting the earth or that astronauts have gone into space. He claims that all photos and videos taken from space are faked. For example, Dubay says that the famous photograph of the earth taken by the Apollo 17 astronauts is a computer-generated image. Of course, this line of argumentation automatically requires belief that the Apollo moon landings were hoaxes. However, there are good reasons to believe that we really did land on the moon during the Apollo program.

Christians who want to entertain this nonsense ought to know that during his six-month stay on the International Space Station in 2006, astronaut Jeffrey Williams photographed the earth more than any astronaut in history. Some of Williams' photos are found in his book, *The Work of His Hands: A View of God's Creation from Space*. Many of the photos show that the earth is spherical. It ought to be apparent from the book's title that Williams is a Christian, and the book's content makes it abundantly clear. Hence, to doubt that the earth is spherical or that astronauts have gone into space is to accuse a Christian brother of perpetuating a tremendous lie.

But Williams is not the only Christian to have gone into space: Jim Irwin and Charles Duke were among the 12 men who walked on the moon. Recently, I asked Brigadier General Charles Duke to respond to those who think that the earth is flat and those who think that NASA faked the Apollo moon landing. This is what he wrote back to me:

> I was the lunar module pilot on the Apollo 16 mission to the moon. We launched from KSC (Kennedy Space Center) in Florida on April 16, 1972. We left earth orbit for our three day trip to the moon about three hours later. As we maneuvered our spacecraft to dock with our lunar module, the earth came into view about 20,000 miles away. It was an awesome sight. ... it [was] obviously a sphere and not a flat circle. As we journeyed to the moon, we would look out our windows and see a smaller earth, and each time we would see different landmasses, so it was obviously rotating on its axis.
>
> Some people are questioning the fact that we landed on the moon, alleging that it is a big hoax. Well, we did land on the moon six times, and the

evidences are overwhelming. If we faked the landing, why did we fake it six times? One needs only to look at the photos from the Lunar Reconnaissance Orbiter from my mission. The photos of our landing site shows the descent stage, the lunar rover, the experiments package, and the tracks we left on the moon. Every landing site has similar evidence. There are many other proofs that conclusively show that Apollo actually landed on the moon six times.

Again, Christians who think that the earth is flat or that men never set foot on the moon are effectively accusing several Christian brothers of lying about one of the biggest things that ever happened in their lives. Are professing Christians among the Apollo moon-landing-deniers prepared to make this accusation?

"Support" for a Flat Earth

Let us discuss some of the more frequent claims that supposedly prove that the earth is flat rather than spherical. Most of the supposed evidences are negative; that is, they are attempts to show that the earth is not spherical. However, at least one, the Bedford Level Experiment, is positive, a direct attempt to show that the earth is flat. In 1838, Samuel Birley Rowbotham claimed to have conducted an experiment on the Old Bedford River on the Bedford Level near Norfolk, England. The Bedford Level is a six-mile stretch of the Old Bedford River that is straight, allowing an uninterrupted view along the six miles. Furthermore, there is no gradient there, so that portion of the river amounts to a slow-flowing drainage canal. If the earth is curved, then the drop from one end to the other is about 24 ft (7 m). That is, if one were to use a telescope at water level to view along the water on one end of the Bedford Level, a mast or pole 24 ft (7 m) high on the other end would not be visible.

Rowbotham waded into the river and used a telescope held 8 in (20 cm) above the water to observe a rowboat with a 5 ft (1.5 m) high mast row away. Rowbotham claimed that he could see the mast when it was 6 mi (10 km) away, even though the spherical earth required that the top of the mast be about 11 ft (3 m) below his horizon (as viewed from 8 in (20 cm) above the water). Rowbotham concluded that the earth must be flat—or it is more likely that he already thought this and this experiment proved his thesis, at least to his satisfaction. Rowbotham, using a pseudonym, published his results in a pamphlet titled *Zetetic Astronomy* in 1849, which he expanded into a book in 1865.

Most people ignored Rowbotham's work. However, in 1870 John Hampden, another flat-earth proponent, offered a wager of a hefty sum to anyone who could demonstrate a convex curvature of a large body of water, as a spherical earth would require. The famous Alfred Russell Wallace took the challenge. Apparently unaware of Rowbotham's result, Wallace altered the technique a bit. He placed two identical objects at different locations along the Bedford Level. Wallace examined either object from a telescope mounted on a bridge. He found that the nearer object appeared higher than the more distant one, consistent with the results predicted by a spherical earth. Why the difference? The density of air decreases with increasing height. Because this causes a slight change in the index of refraction in air, rays of light passing close to the earth's surface are bent downward. As can be seen in the following figure, this makes distant objects appear higher than they actually are (Figure A.1).

Incidentally, this well-understood effect causes the sun to appear to rise about two minutes earlier than it actually does. A temperature inversion, where the temperature increases with height, is common at low heights along the Bedford Level and other bodies of water. Temperature inversions accentuate refraction. If the rate of increase of air temperature with height is great enough, a temperature inversion can even cause objects far in the distance to appear

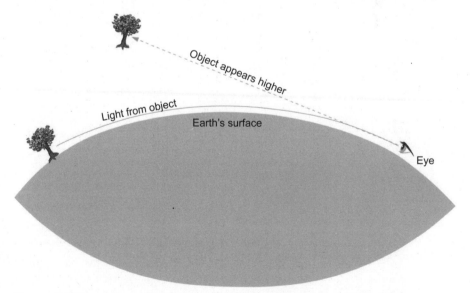

Figure A.1. Illustration of refraction of light passing close to earth's surface.

above the horizon. In 1896, Ulysses Grant Morrow conducted a similar experiment on the Old Illinois Drainage Canal under these conditions and found results consistent with the earth being curved concavely (there are, strangely enough, people who think that the earth's surface is the inside of a shell). Apparently, Wallace was aware of these effects, while Rowbotham was not. This is what prompted Wallace to conduct his experiment high enough above the water to eliminate the major contribution of refraction due to a temperature inversion at low height.

Those who promote the flat earth often mention the Bedford Level Experiment as proof that the earth is flat. They seem to think that Rowbotham's 1838 experiment settled the matter for all time. They are willfully ignorant that the experiment has been repeated many times since 1838. When those experiments are properly conducted to minimize the effect of refraction, they are consistent with a spherical earth.

What other "evidence" for a flat earth has been set forth? Some Internet videos promoting the flat earth show a time-lapse film of the midnight sun. The sun appears to move rightward along the horizon, slowly bobbing up and down once each day. The claim is made that the midnight sun is visible anywhere north of the Arctic Circle (around 66.6° north latitude), but that if the earth were spherical, the midnight sun would be visible only at the North Pole. The following figure shows the correct situation (Figure A.2).

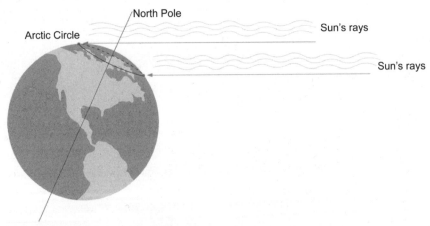

Figure A.2. Illustration of midnight sun on spherical earth.

On the summer solstice, the earth's Northern Hemisphere has its maximum tilt toward the sun. Consider an observer on the Arctic Circle. At point A, it is noon, and the sun is as high in the sky as it can be, nearly 47°. To an observer facing the sun with the North Pole to his back, the sun would appear in the southern part of the sky. However, 12 hours later the earth's rotation will take the observer to point B. This will be at midnight. As you can see, the sun's rays pass over the North Pole and reach point B tangent to the earth's surface. The sun's rays being tangent to the earth's surface means that the sun is on the horizon. Since the observer must face the North Pole to view the sun, the sun is in the northern part of the sky.

On the edge of the Arctic Circle, the midnight sun is visible only on the summer solstice. At higher latitudes, the midnight sun is visible for more days. At the earth's North Pole, the sun is above the horizon for six months. The sun does not appear to bob up and down each day at the North Pole. Instead, the sun appears to circle each day at about the same altitude. Actually, the sun rises on the vernal equinox and slowly gains altitude until the summer solstice, whereupon the sun slowly descends again until it sets on the autumnal equinox. The sun's maximum altitude on the summer solstice is 23.4°.

There is an irony here. While supporters of the flat earth falsely claim that the midnight sun on the Arctic Circle cannot happen if the earth is spherical, it is the flat earth that has difficulty explaining the midnight sun. Most flat-earth models have the North Pole at the center of a disk-shaped earth, as in the following illustration (Figure A.3).

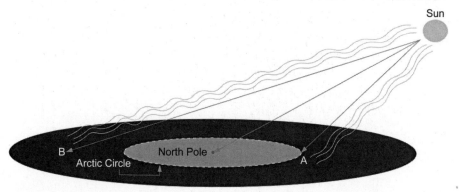

Figure A.3. Illustration of the impossibility of midnight sun on flat earth.

Suppose that the midnight sun is visible at the North Pole as well as on the Arctic Circle. This is indicated by lines from the sun to the North Pole and at point A on the Arctic Circle. Notice that on a flat earth, we can draw a line from the sun to any point on the earth not within the Arctic Circle (such as point C). Hence, if the earth were flat, the midnight sun must be visible everywhere, not just within the Arctic Circle. Because this clearly is not the case, the earth must not be flat.

Some of the flat-earth promotional videos that deal with the midnight sun show the sun orbiting each day around the earth's North Pole. Mysteriously, there is a shadow on the earth on the other side of its North Pole opposite the sun. As the sun orbits the North Pole, so does the shadow. Apparently, the shadow indicates where it is night on the earth. However, because the sun clearly is above the horizon for locations in that shadow, it ought to be day there. The origin of this shadow-producing night is never explained. Furthermore, since the sun clearly is above the horizon for the entire flat earth, it ought to be day everywhere on the earth. This, too, is not explained.

Another claim made against the spherical earth is that if the earth were a spinning globe that orbited the sun each year, the earth's spin axis would not stay aligned with the North Star. This is because, as we shift from one side of the earth's orbit to the other, our perspective changes, as can be seen in Figure A.4.

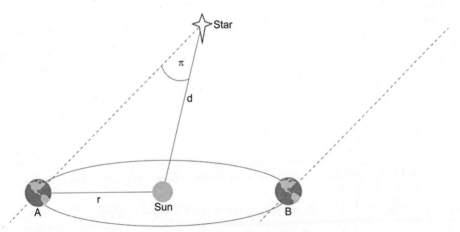

Figure A.4. Illustration of stellar parallax as assumed by flat-earth proponents.

If the earth's axis were aligned with the North Star at point A, then the earth would not align with the North Star six months later when the earth arrives at point B. This effect is well enough known to warrant a name: parallax. At least one of the videos gives what astronomers think is the distance to the North Star (four quadrillion kilometers, but it's actually about twice that distance) and the radius of the earth's orbit (95,000,000 mi [150,000,000 km]). We can use these numbers to find how much the parallax angle, π, is. As we shall see, the angle π is a small angle, so we can use the small angle approximation. If an angle is small, we can express the angle, in radian measure, as the ratio of the baseline to one of the other sides. The baseline is the earth's orbital radius, r, and the other side is the distance to the North Star, d. That is,

$$\pi = r/d = (150 \text{ million km})/(2 \text{ quadrillion km}) = 7.5 \times 10^{-8} \text{ radians.}$$

To convert this to degrees, we must multiply by 57.3. After doing this, the angle is 4.3×10^{-6}°, or a little more than four millionths of a degree. That is the apparent diameter of a dime when viewed 150 mi (250 km) away. The parallax is defined as half the total shift, so the total shift that we would see would be twice this amount, but remember, the distance given in the video is about half the true value. Obviously, this is a very small angle, far too small for our eyes to notice. Therefore, this supposed proof that the earth is flat is specious.

Supposed Biblical Support for a Flat Earth

Proponents of the flat earth who profess to be Christians commonly assert that the Bible indicates the earth is flat and list biblical passages which supposedly support this idea. It is interesting that many of the passages given in support are the same passages that skeptics list in an attempt to discredit the Bible for allegedly teaching that the earth is flat. It causes one to wonder if at least some of the flat-earth promoters who claim to be Christians actually are skeptics engaging in a stealth campaign against the Bible. Here we will examine some of these passages which supposedly indicate that the earth is flat. Unfortunately, many of the presentations of biblical passages that supposedly support a flat earth merely quote verses with no comment or explanation. Instead, there is usually an invitation to read all the passages listed and then to ponder what these verses mean in totality. With little or no discussion, one is left to conjecture just how the authors intended to interpret these verses so that they support a flat earth. Rather than addressing the earth's shape, some of the listed biblical passages apparently are intended to support concepts that are related to the snow globe earth model. Besides the earth being flat in the snow globe earth model, there are at least two related ideas: that the earth is stationary, and that it is surrounded by a hard sphere on which astronomical bodies are affixed. So now let us consider the verses that supposedly support these three claims, that the earth is stationary, that there is a solid sphere surrounding the earth, and that the earth is flat.

Is the Earth Stationary?

Let us first examine some of the verses that supposedly teach geocentrism, with its accompanying belief that the earth does not move. Chapter 13 of my previous book, *The Created Cosmos: What the Bible Reveals about Astronomy*, discussed the flat earth and geocentrism in some detail; what I write here must necessarily repeat some of that material. However, the version of geocentrism addressed there was the Tychonic model. In the Tychonic model, the earth is stationary and is orbited by the moon and sun. However, the other planets orbit the sun, so that the motion of the planets with respect to the earth is a combination of their motion around the sun and the sun's motion around the earth. In some sense, this is a heliocentric (sun-centered) model, and it amounts to a coordinate transformation from the sun to the earth. Promoters of the flat earth do not discuss the motion of the planets, so it is not clear what sort of model they have for the planets' motion (the lack of discussion suggests that they have not even considered this question). At any rate, it is clear that those who believe the Tychonic model want nothing to do with those who promote a flat earth.

Some supporters of the Tychonic model quote Bible verses that speak of the sun rising or setting. Treating these verses in a hyper-literal manner, they conclude that earth does not spin. Hence, the heliocentric model with a spinning earth must be false.[7] How do we who believe in the authority of Scripture and a spinning earth respond to this? The earth spins, causing the sun to *appear* to rise and set. Relative and absolute motion are tricky concepts.

---

[7] I must emphasize that supporters of the Tychonic model are split. While all of them believe that the earth does not move through space, some believe that the earth rotates on its axis each day, while others do not. Obviously, the ones that believe that the earth does not rotate are the ones who argue that verses which speak of sunrise and sunset show that the Bible is geocentric.

In some sense we can say that the sun *does* rise each day, move across the sky, and then set. To be absolutely literal about it, we properly ought to say that the sun *appears* to rise, *appears* to move across the sky, and *appears* to set. This is a bit clunky, and no one speaks this way. We are totally comfortable with people speaking in a phenomenological sense, that is, in the manner describing the way things appear to happen. Professional astronomers, such as myself, nearly always talk and write in this way of the sun, moon, and stars rising and setting, yet no one accuses us of denying the earth's rotation. Why would the Bible be any different? In many respects, this is a moot point because the flat-earth geocentrists never mention these verses. It may be because, as already discussed, sunrise and sunset are difficult concepts to explain for those who believe in a flat earth.

One verse that flat-earth supporters use to show that the earth does not move is Joshua 10:13, wherein Joshua commanded the sun and moon to stand still.[8] Certainly, it is argued, this must mean that it is the sun and moon which move, or else Joshua would have commanded the earth to stop rotating. As just argued with the rising and setting of the sun, even professional astronomers often speak of the sun, moon, and stars moving in the sky—even though we know that it is the rotation of the earth that causes the motion. In that sense, it is quite reasonable for Joshua to have used the words that he did.

A common verse appealed to by geocentrists of all types, including flat-earthers, is 1 Chronicles 16:30, which reads,

> Tremble before him, all the earth;
> yes, the world is established; it shall never be moved.

Surely, they reason, if this verse means anything, it means that the earth does not move. Or does it? What is the context of this verse? In 1 Chronicles 15, David brought the Ark of the Covenant to Jerusalem, and in 1 Chronicles 16:1–7, David brought the Ark into a tent and presented offerings before it. The narrative continues with a song of thanksgiving in 1 Chronicles 16:8–36, which includes the verse in question. Being a song, it amounts to poetry. Indeed, portions of this song are repeated in Psalm 96, 98, 100, and 106. For instance, the relevant portion of 1 Chronicles 16:30 appears word for word in

---

[8] For a more detailed discussion of Joshua's long day, see Chapter 6 of *The Created Cosmos: What the Bible Reveals about Astronomy*.

Psalm 96:10. (Psalm 96:10 is likewise included in most lists of verses allegedly supporting a stationary earth.) However, a hallmark of poetry is imagery and figurative language. For instance, 1 Chronicles 16:31, which immediately follows 1 Chronicles 16:30, states,

> Let the heavens be glad, and let the earth rejoice,
>   and let them say among the nations, "The LORD reigns!"

Notice here that the heavens and earth are personified, for the heavens are described as being glad and the earth is commanded to rejoice. It is doubtful that even the most literal of literalists (including flat-earthers) would insist that the heavens can feel emotions or that the earth can speak. They would agree that these are poetic statements. Yet flat-earthers insist that 1 Chronicles 16:30 must mean that the earth does not move in some literal sense. The phrase "the world is established; it shall never be moved" from 1 Chronicles 16:30 and Psalm 96:10 essentially says the same thing two different ways. In these verses, the earth not moving is not to be understood as a reference to its lack of motion, but to the fact that God created the earth to endure—God created *and* continually sustains the earth. The modern geocentrists are claiming that these verses mean something other than what the authors intended them to mean in their respective literary and historical contexts. Another verse used to argue that the Bible teaches an immovable earth is Psalm 93:1, which reads,

> The LORD reigns; he is robed in majesty;
>   the LORD is robed; he has put on strength as his belt.
> Yes, the world is established; it shall never be moved.

The relevant phrase, "the world is established; it shall never be moved" is the exact same wording of 1 Chronicles 16:30 and Psalm 96:10. This is because the Hebrew wording underlying the translation is likewise the same.

In a similar manner, flat-earthers reference verses that mention the foundation(s) of the earth (e.g., Job 38:4; Psalm 102:25, 104:6; Isaiah 48:13) or the pillars of the earth (e.g., 1 Samuel 2:8; Job 9:6; Psalm 75:3). If taken in a rigidly literal fashion, the earth having a foundation or pillars would suggest that the earth rests upon something. What is the earth resting upon? What is its foundation? If one views these verses strictly in a literal sense, the earth itself cannot be the foundation or pillars. Strangely, though, flat-earthers also quote

Job 26:7:

> He stretches out the north over the void
> and hangs the earth on nothing.

This verse clearly indicates that the earth is supported by nothing. One might claim that this verse merely means that the earth is not *suspended* by anything (such as a cable), but that the earth could be supported by a foundation or pillars below. But wouldn't that mean that one could now understand Job 26:7 (interpreting Scripture in terms of Scripture) as, "He stretches out the north over the void and hangs the earth on nothing (but it is supported below)"? Clearly, such a reading would entirely gut the majesty and power of God that Job 26:7 conveys.

The problem is that most, if not all, of the cited verses come from poetic passages. Poetry, including ancient Hebrew poetry, contains much imagery. By its very nature, imagery employs figurative, symbolic allusions. That is to say, there are many aspects of poetry that clearly are non-literal. This does not mean that *everything* in poetry is non-literal, because then poetry would be meaningless. Hence, one must exercise some caution in exegeting poetic passages in the Bible. For instance, the book of Job is classed with the poetic books of the Old Testament, because it clearly is written in a poetic style. However, that does not mean that the man Job merely was a poetic device. Rather, Job was a real, breathing human being who endured hardship. It is just that his story of suffering is told (mostly) in a poetic style.[9] Often, conservative Christians are accused of believing that every word in the Bible is to be taken literally. This clearly is not true. Unfortunately, the hyper-literal manner of interpretation employed by those who claim that the Bible teaches a flat earth provides fodder for this false accusation.

Is the Earth Surrounded by a Hard, Transparent Sphere?

The snow globe earth model features a flat earth with a hard, transparent dome above. Many times, presentations of this supposed biblical cosmology feature a figure illustrating this model, as the accompanying figure here shows (Figure

---

[9] Outside of the traditional poetic books there are often some poetic passages present. Some critical scholars claim that poetic elements of the first few chapters of Genesis override all other considerations and thus that none of the events described therein actually happened. That is, there was no literal Adam, no Garden of Eden, no temptation, and no Flood in Noah's day. This is nonsense, because there are many hallmarks of historical narrative found in the book of Genesis.

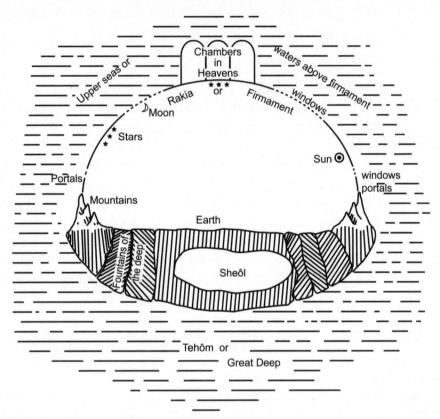

Figure A.5. Illustration of "snow globe earth". (G.L. Robinson, *Leaders of Israel*. New York: Association Press, 2)

A.5). Where do flat-earthers get the idea that the Bible teaches this? Certainly, such illustrations were never part of the inspired canon of Scripture. Many of the verses cited in support of this concept involve the word *firmament*. This word occurs 17 times in the King James Version of the Bible, though more modern translations often translate the underlying Hebrew word as *expanse*. The Hebrew word is *rāqîaʿ*. Its exact meaning is not readily apparent, which has led to confusion and uncertainty. A major part of the problem is that the word occurs so infrequently, with more than half of the instances of its use in Genesis 1 alone.

The noun *rāqîaʿ* derives from a verb that means to beat, press, or stamp out. A good illustration of this verb would be in the beating or rolling out of gold into thin sheets. Gold is very malleable, which allows craftsmen to spread gold out into very thin foil that they can apply to surfaces to give the impression of a solid gold object. It would appear that the intended property of this thing called the *rāqîaʿ* is that it is spread out, or expanded. That is why many modern English translations render it as *expanse*. As explained in Chapter 2 of *The Created Cosmos: What the Bible Reveals about Astronomy*, the unfortunate translation of *rāqîaʿ* as *firmament* goes back to the Septuagint more than 2000 years ago in an attempt to wed the Genesis creation account to the Hellenistic cosmology of that day. That cosmology included a solid, transparent dome over the earth to which the astronomical bodies were attached. This was similar to the cosmologies of ancient cultures that surrounded ancient Israel, so many Bible scholars have incorrectly assumed that the Genesis creation account must reflect the thinking of that era of antiquity. Hence, some of these Bible scholars have constructed diagrams of what they think the ancient Hebrew cosmology must have been. This is the view that modern flat-earthers have embraced.

However, as shown in Chapter Two of *The Created Cosmos: What the Bible Reveals about Astronomy*, this view is almost certainly wrong. Rather, the *rāqîaʿ* of Genesis 1 probably refers to what we would call space today, with perhaps at least part, if not all, of the earth's atmosphere included as well. Hence, there is no biblical reason compelling us to believe in the snow globe earth.

Does the Bible Say the Earth Is Flat?

The phrase "ends of the earth" appears 28 times in the King James Version (e.g., Job: 28:24; Isaiah 41:9; Jeremiah 16:19), though more modern translations may vary on the wording of some of the occurrences. Supporters of a flat earth insist that these verses must refer to a physical edge to the earth, requiring that the earth be flat. Presumably the edge of the flat earth is where the transparent globe surrounding the earth intersects the earth, beyond the ice wall that we call Antarctica. Actually, the phrase "ends of the earth" is idiomatic, and it refers to the remotest parts of the earth. The phrase "four corners of the earth" appears three times in the Bible. Many skeptics claim that this must refer to a flat, square earth, thus proving that the Bible teaches a flat earth. There are some examples of flat-earth cosmologies from the ancient world, but they always consisted of a flat, *round* earth. A circle was considered a much more perfect shape than a square, so none of the flat-earth cosmologies were square. If a square flat earth were the cosmology of the Bible, then it would have been at odds with every other ancient flat-earth cosmology. The one occurrence of "the four corners of the earth" in the Old Testament, Isaiah 11:12, should be taken in the same manner as other Old Testament passages mentioning the "ends of the earth." It is likewise an idiomatic expression. The two New Testament occurrences of "the four corners of the earth" are in the book of Revelation. Revelation 7:1 speaks of four angels standing on the four corners of the earth and restraining the four winds of the earth. The four winds obviously refer to the four directions of the wind: north, south, east, and west. This repetition ("four angels...four corners...four winds") makes it clear that this is an idiom referring to the four compass directions. The phrase "four corners of the earth" from Revelation 20:7 also is idiomatic, referring to the four directions. Since most flat-earth enthusiasts today believe that the earth is flat but round, they generally do not mention these verses, though a few do.

One of the more bizarre biblical passages that flat-earth supporters claim proves their point is Daniel 4:10–11, which reads:

> The visions of my head as I lay in bed were these: I saw, and behold, a tree in the midst of the earth, and its height was great. The tree grew and became strong and its top reached to heaven, and it was visible to the end of the whole earth.

Presumably, this passage teaches a flat earth, because only on a flat earth could the top of the tree be visible from the entire earth. Alternately, perhaps only in a snow globe earth could the top of a tree reach to heaven. There are many problems, however, with claiming that this passage indicates that the Bible teaches the earth is flat. For instance, even flat-earth advocates agree that the tallest mountains are not visible over the entire earth—the angular height of the mountains is exceedingly small when viewed from hundreds, or even thousands, of miles. To be readily visible over the entire flat earth, a tree would have to be hundreds, if not thousands, of miles high. Do flat-earth supporters think that this tree actually existed? No, for this tree appeared in a *dream* that King Nebuchadnezzar had one night. If one reads Daniel 4 in its entirety, one will readily see that Nebuchadnezzar described what he experienced as a dream (verses 5, 6, 7, 8, 9, 18, and 19), and Daniel called it a dream too (verses 19). Four times Nebuchadnezzar called his dream a vision (the phrase "visions in my head" occurs in verses 5, 10, 13, and "visions of my dream" appears in verse 9). In verses 8–18, Nebuchadnezzar recounted his dream to Daniel and requested that Daniel interpret the dream. Daniel interpreted the dream in verses 19–27, in which Daniel said that Nebuchadnezzar was the tree in the dream (verse 22). Daniel told Nebuchadnezzar that because of his pride, he would lose his mind for seven years. The fulfillment came a year later (verses 28–33), but after seven more years Nebuchadnezzar was restored to his mind and his kingdom (verses 34–37).

Key to properly understanding this text is that Daniel metaphorically identified Nebuchadnezzar with the tree. That is, there was no literal tree, because it appeared in a dream in which the tree *represented* Nebuchadnezzar. We have no idea whether Nebuchadnezzar believed in a flat earth and thus understood his dream within that cosmology. It really doesn't matter, because this dream does not address cosmology, and even if it did, it would not amount to an endorsement of any particular type of cosmology on God's part. Rather, it would be the statement from Nebuchadnezzar, who likely was a pagan at the time, concerning his cosmology, not revelation from God about what particular cosmology were true.

Supporters of the flat earth probably have similar reasoning in mind when they cite Matthew 4:8–9, which reads,

> Again, the devil took him to a very high mountain and showed him all the kingdoms of the world and their glory. And he said him, "All these I will give you, if you will fall down and worship me."

This account concerns the temptation of Jesus after His 40-day fast (Matthew 4:1–11). Presumably, those who believe that the earth is flat think that it would be impossible to view all the kingdoms of the world from a high mountain if the earth were spherical, but it would be possible if the earth were flat. However, if the flat-earthers are correct on this matter, then shouldn't there be a mountain on the earth from which one literally can see all the kingdoms on earth? If so, where is this mountain? Clearly, no such mountain exists.[10]

How do we properly understand this second temptation? Some clues may come from the parallel account in Luke 4:1–13. Verse 5 in the King James Version

---

[10] One could argue that this exceedingly high mountain exists beyond the ice wall in Antarctica. However, if all the earth's kingdoms were visible from such a mountain, conversely that mountain must be visible from all the earth's kingdoms. Why hasn't anyone spotted this towering mountain?

includes the description of the Devil taking Jesus up to a high mountain to show Him all the kingdoms of the world, but most modern translations of Luke's account omit the mention of a high mountain, stating simply that the Devil "took Jesus up." This is because while the phrase concerning a high mountain exists in the *Textus Receptus* of Luke's account of the temptation of Jesus, earlier, more reliable manuscripts do not include this phrase. This probably is because a later copyist, being familiar with the language of Matthew's account, inserted the additional words into Luke's account. More revealing, Luke's gospel adds a mention of a time element by stating that the Devil showed Jesus all the kingdoms of the world "in a moment of time." A more literal rendering into English would be "in a point of time." That is, the Devil showed Jesus all the world's kingdoms virtually instantly. This would not be possible if indeed the Devil literally took Jesus to a high mountain and literally pointed out each of the world's kingdoms. That is not to say that the temptation did not take place; it did. Rather, certain elements of the temptation may not be as literal as some would propose. For instance, this portion of the temptation may have been a vision that the Devil shared with Jesus.

Yet another sort of related reasoning apparently occurs when those who believe that the earth is flat list Revelation 1:7 as support. That verse refers to the return of the Lord Jesus Christ, and reads,

> Behold, he is coming with the clouds, and every eye will see him, even those who pierced him, and all tribes of the earth will wail on account of him. Even so. Amen.

Once again, the argument appears to be that it would be impossible for every eye to see Jesus' return if the earth were spherical, but it would be possible if the earth were flat. Again, strict hyper-literalism fails the test. Notice that "those who pierced" Jesus are included among those who will witness the Lord's return. Who pierced Jesus' side? According to John 19:31–37, the soldiers who crucified Jesus literally pierced His side (and, more specifically, according to verse 34 it was *one* of the soldiers who actually thrust the spear into Jesus' side). In his Crucifixion account, the Apostle John quoted Psalm 34:20, which states that "not one of his bones would be broken," as well as Zechariah 12:10, which says that "they will look on him whom they pierced." The Apostle John quoted his own gospel and Zechariah 12:10 in Revelation 1:7. But wouldn't those who pierced Jesus' side be at *least* 2000 years old when they will witness the Lord's return? Obviously, this cannot be true in a rigidly literal sense. However, since Jesus died for the sins of all mankind, then in a metaphorical sense, we all are responsible for Jesus' Crucifixion and hence also are responsible for piercing His side. Clearly, there are certain elements of this verse that are not absolutely literal.

Coming back to the question at hand, is it possible for *every eye* to see the Lord's return, even on a spherical earth? Of course. To doubt this would be to limit God's power. Christ's return will be a display of God's power in a *miraculous* way. Therefore, this verse hardly teaches that the earth is flat, and those who insist that it does obviously have approached it with an agenda or, at the very least, with several hidden assumptions.

Conclusion

Many of the arguments put forth by Dubay and others for a flat earth are so poor that one has to wonder how serious these people must be.

Are these people who believe in a flat earth for real? It's hard to say. They could be well-intentioned but seriously misguided people. Or they could be attempting to discredit the Bible and Christianity. If the latter, their approach probably is, "If you think that the Bible is literally true, then I'll show you just how literally true the Bible is!" But this is a false dichotomy. Christians who believe in the inspiration of the Bible and have a high regard for the authority of Scripture usually don't say that the Bible is literally true. Rather, they understand that the Bible is true because it is inspired by God. As such, it is authoritative on all matters and is reliable. The Bible contains imagery and poetry. However, those passages are easy to identify. When it comes down to the sorts of questions that matter here (such as "Did God create the world?"), the Bible must be read and understood historically and grammatically. That is, historical narrative does not lead to symbolic interpretation. Hence, the creation account is literally true.

At least some of the people behind this upsurge in the flat-earth movement may be lampooning the creation movement. As such, they clearly are no friends of the church; rather, they oppose Christ and His kingdom. Christians therefore need to be very discerning about the teachings of those who espouse a flat earth.

Appendix B

The August 21, 2017 Total Solar Eclipse

Due to several circumstances, this book was delayed a few weeks before going to press. The final hitch was that we discovered a problem with the pagination. A book such as this is printed in sixteen-page sections, requiring that the number of pages be a multiple of sixteen. Then there is the section in the back with color photos. It requires special paper, and it too must be in eight-page segments. The color images that I originally intended for the book took up nine pages. So, what were we to do, eliminate one page of color images or add seven more pages of them? Plus, the text of the book ran three pages over a multiple of sixteen, which would leave five blank pages before the color section. How could we solve this problem?

As it turns out, this was providential because we discovered the problem on August 22, 2017. The first total solar eclipse in the continental United States in 38 years was just the day before. In fact, I had traveled to Oregon to watch the eclipse, and I found out about this problem as I was boarding a plane in Portland to return home to Kentucky. I realized that we could fill those seven pages in the color section with photos that I and others had just taken. Furthermore, the appendix you are now reading could take up the additional five pages of text. Consequently, this may be the first book to appear in print in the United States with color images of this eclipse (see Appendix C).

I had traveled to Oregon to view the eclipse near Warm Springs, Oregon. My trip was organized by Marianne Pike, who arranged speaking engagements with the Design Science Association (DSA), the Institute for Creation Science, and Oregon City Evangelical Church the weekend before. Marianne also planned the DSA field trip to view the eclipse. There were more than 40 people on the field trip. Marianne's sister, Beki Pike, organized videography of the trip, and I expect her to produce a DVD of the experience. I wish to express my deepest gratitude to Marianne and Becki for making this trip possible.

No one in our group lived in the eclipse path, so we used a bus to transport people from the Portland area across the Cascade Mountains. We wanted to be east of the Cascades to avoid morning clouds that sometimes plague the Oregon Coast and the Willamette Valley. With concerns about huge throngs of people having the same idea and traveling into the eclipse path that day, we decided to leave very early. So, we departed from Sandy, Oregon at 3:00 AM to ensure we would reach our destination in plenty of time. For an eclipse, it's much better to be much too early than to be a little too late. We arrived at our chosen spot before sunrise. Dawn had started, so it wasn't totally dark, but I got out of the bus to look at the stars that were visible. The location to view the eclipse was well chosen, because it was on private land, which provided privacy for our group. It was on a hill in a pasture with horses. There was a good view to the east. When totality ended, we watched the moon's shadow depart that direction.

A total solar eclipse is one of those things that must be experienced truly to understand its beauty and wonder. My first eclipse was the one in 1979, and I was completely unprepared for what I saw. I expected my second eclipse, in 2017, to be much the same. It wasn't. I've heard people who have been to many eclipses say that each one is very different, but I didn't understand. Now I do. It is a gross understatement to say that a total solar eclipse is the most remarkable thing I have every experienced.

If the sky is very clear, by the time the sun is 70-80% covered by the moon, it is noticeable that there is less light than normal. The amount of light is like that on a day with thin clouds. On slightly overcast days, shadows are at best fuzzy, if visible at all. However, when the obscuration is due to the sun being partially eclipsed, shadows are more distinct than ever. This adds an eerie effect that grows as the sun is more blocked by the moon. It builds to a crescendo when everything rapidly goes dark.

Nearly everyone knows that it gets dark for a few minutes in a total solar eclipse, but there is far more. The darkness isn't like midnight. Rather, it's more like dusk. With the sun shining a few miles away in every direction, the sky is brighter near the horizon. It is tinged with an orange glow, like a 360-degree sunset. The brighter stars and planets often are visible. In 1979, I didn't see any stars or planets. That probably was due to the thin clouds that covered the sky. At this eclipse, we had a few thin clouds, but I was able easily to see Venus and the stars Sirius and Procyon. I briefly looked for other stars, but I couldn't see any. I had more important things to look for than stars that might not even be visible!

Totality is difficult to describe. At the 1979 eclipse, I was struck by the many large prominences, blood-red loops extending outward around the edge of the sun. I expected to see these again in 2017, but there were just a few modest ones. However, the solar corona this time was far more impressive. It extended out several radii of the sun. Through the corona there were streamers that outlined the sun's magnetic field. No photograph truly can capture what the eye sees in the corona. This is because the eye and camera operate very differently. A camera records light linearly. That is, if light level is doubled, the camera response is doubled. But the eye records light logarithmically. This means that if light intensity is doubled, the response in our eye may be 20% more. This gives our eyes far greater dynamic range than cameras have. Simultaneously we can see very feeble and very bright light sources. But cameras can record either bright light levels with a shorter exposure time or faint light levels with longer exposure times. A short exposure reveals the inner corona but not the outer corona. Increasing the exposure time reveals the outer corona, but the inner corona is overexposed. Neither extreme in exposure time can capture the middle corona well. However, the eye can see all the corona at one time. The corona is the most beautiful thing I have ever seen. And, again, no photograph even begins to capture how it looks to the eye.

The sudden darkness has a profound effect on animals. Less than a minute before totality, a bat flew over our location. Apparently, it was roused from its sleep, thinking that it was dusk, signaling the start of a night of feeding on insects. We wondered if it was confused just a few moments later when dawn broke, beckoning the bat to return to its slumber. As totality ended, horses just down the hill from our position began to run about and neigh. Perhaps that was their morning routine. I've heard that animals have no sense of time, so perhaps they thought nothing of such a brief night.

People certainly do have a sense of time. We also like to think that a total solar eclipse doesn't affect us so much. That decidedly is not the case. The video of our 2017 eclipse experience reveals that I shouted at the beginning of the eclipse. It must have been spontaneous because I have no recollection of that. In 1979, we ran a tape recorder during totality. While other people off in the distance were hooting and hollering, I didn't say two words. I guess I was dumb struck. This demonstrates what a moving experience a total solar eclipse is. We all react in different ways, and often we react differently at different times. I was curious how I would react this time. It makes me wonder what my reactions will be during future total solar eclipses (eclipses are habit-forming, so I intend to do this again).

What is the takeaway from a total solar eclipse? It is the most incredible, indescribable thing one can experience. Someone who is not moved by the experience must have something terribly wrong with them. As discussed on p. 133, there are unusual circumstances that cause total solar eclipses to be both rare and spectacular. The sun is 400 times larger than the moon, but the sun is 400 times farther away. If the circumstances were any different, total solar eclipses would either not happen at all or such eclipses would be far more frequent and markedly less spectacular. This is the only planet in the solar system where such rare and spectacular total solar eclipses happen, and it is the only planet where such a thing matters because humans, the only creatures that can appreciate total solar eclipses, inhabit the earth. One could insist that this is a matter of coincidence, but how many coincidences is one allowed to have? The earth has many unique properties that make life on it possible (just a few of which are discussed in this book). Are they coincidences too?

Of course, unlike those other amazing properties of earth, total solar eclipses are not necessary for our existence. However, there is much more to human life than just existence. According to the Bible, man was created in God's image. Being in God's image is a complex topic. Among other things, it certainly includes our intellect, language ability, and social nature, and our ability to have a relationship with God. But it also includes a deep appreciation of beauty and wonder, something that animals appear to lack. God put beauty in much of His Creation, making Him the consummate artist. I think it is that part of God's image within us that gives us appreciation for the beauty in the world around us. Total solar eclipses are so far off the charts in terms of artistry that I must conclude they are a special gift from God. Consequently, for those people who know God, a total solar eclipse is deeply reverential and proclaims God's majesty. I like to think that the dazzling appearance of the solar corona is a mere picture of the glory of God (for instance, Matthew 17:2 records that at His transfiguration, Jesus' face shown like the sun).

The weather was clear over much of the path of totality from Oregon to South Carolina, so the millions of people who took advantage of this rare opportunity were rewarded. Even the coast and Willamette Valley of Oregon were clear, something that one cannot count on early in the morning. There were exceptions because it was cloudy in coastal South Carolina and the Kansas-Missouri border. Consequently, not only did I get some good photographs of the eclipse, but friends scattered across the nation did also. On the pages that follow, you will see photographs that I and some of my friends took during the 2017 eclipse. I hope that you enjoy them nearly as much as we had in taking them.

Appendix C

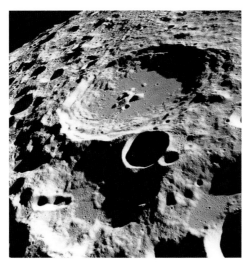

Image A. Determination of relative ages of lunar features (maria and craters). (NASA. Public Domain).

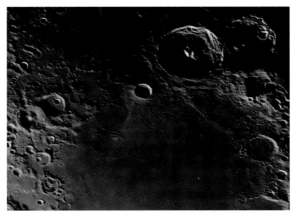

Image B. Photograph of lunar ghost craters. (Photograph: Danny Faulkner).

Image C. Illustration of solar eclipse. (NASA. Public Domain).

Image D. Photograph of Mimas. (JPL. Public Domain).

Image E. Image of sun's photosphere. (Photograph: Danny Faulkner).

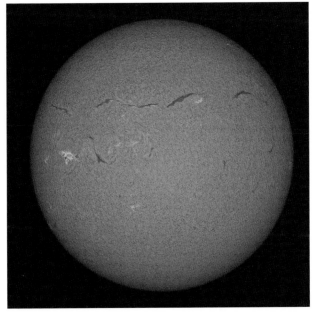

Image F. Image of sun's chromosphere. (Photograph: Glen and Katrina Fountain).

Image G. Photograph of planetary nebula. (JPL. Public Domain).

Image H. Photograph of planetary nebula. (JPL. Public Domain).

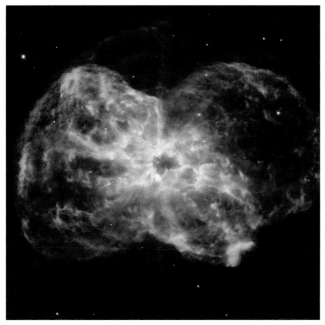

Image I. Photograph of planetary nebula. (JPL. Public Domain).

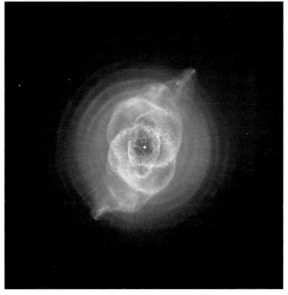

Image J. Photograph of planetary nebula. (JPL. Public Domain).

Image K. Photograph of planetary nebula. (Photograph: Glen and Katrina Fountain).

Image L. Photograph of spiral galaxy. (JPL. Public Domain).

Image M. Photograph of spiral galaxy. (JPL. Public Domain).

Image N. Photograph of spiral galaxy. (JPL. Public Domain).

Image O. Photograph of spiral galaxy. (JPL. Public Domain).

Image P. Photograph of elliptical galaxy. (JPL. Public Domain).

Image Q. Photograph of elliptical galaxy. (JPL. Public Domain).

Image R. Photograph of M51. (JPL. Public Domain).

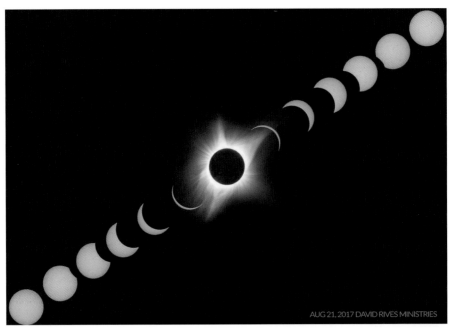

Image S. Eclipse sequence. (Photograph: David Rives Ministries: www.davidrives.com).

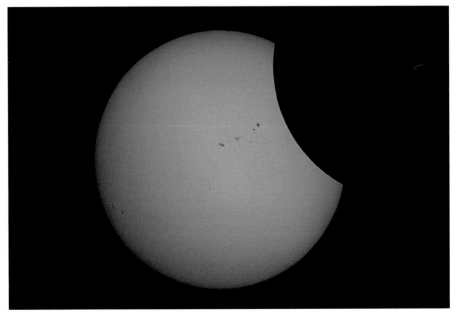

Image T. Partial eclipse. (Photograph: Danny Faulkner).

Image U. Eclipse, diamond ring stage, 2017. (Photograph: Jim and Deb Bonser).

Image V. Total eclipse, 2017. (Photograph: Jim and Deb Bonser).

Image W. Eclipse, totality, 2017. (Photograph: Danny Faulkner).

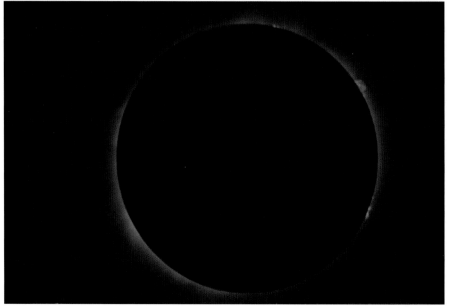

Image X. Eclipse, totality, 2017. (Photograph: Danny Faulkner).

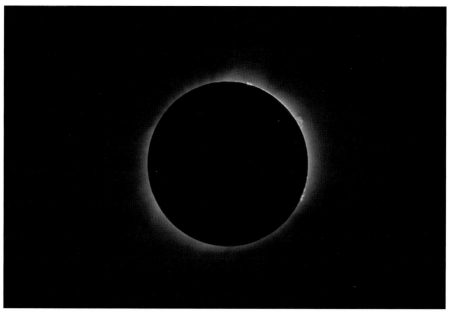

Image Y. The inner corona and prominences. (Photograph: Jim and Deb Bonser).

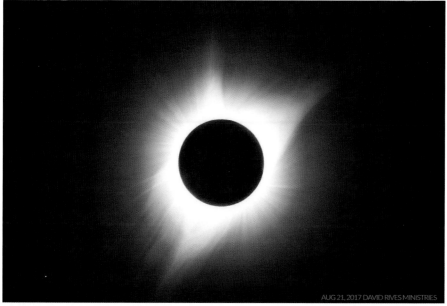

Image Z. Total solar eclipse showing the solar corona, the Star Regulus to the lower left. (Photograph: David Rives Ministries: www.davidrives.com).

Image AA. Lunar Detail. (Photograph: David Rives Ministries: www.davidrives.com).

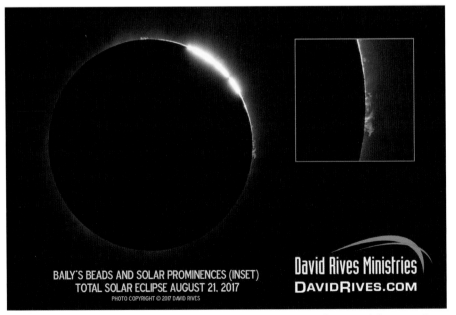

Image BB. Bailey's Beads, with details of the prominences in the inset (Photograph: David Rives Ministries: www.davidrives.com)

Image CC. Eclipse, ground view, 2017. (Photograph: Jebi Koilpillai).

Image DD. The wonder and amazement of the 2017 total solar eclipse shows in Danny Faulkner's face. (Photograph: Jebi Koilpillai)

Subject Index

Scripture Index